Exercises for Emb

Exercises for Embodied Actors: Tools for Physical Actioning builds on the vocabulary of simple action verbs to generate an entire set of practical tools from first read to performance that harnesses modern knowledge about the integration of the mind and the rest of the body.

Including over 50 innovative exercises, the book leads actors through a rigorous examination of their own habits, links those discoveries to creating characters, and offers dozens of exercises to explore in classrooms and with ensembles. The result is a modern toolkit that empowers actors to start from their own unique selves and delivers specific techniques to apply on stage and in front of the camera.

This step-by-step guide can be used by actors working individually or by teachers crafting the arc of a course, ensuring that students explore in physically engaged and dynamic ways at every step of their process.

Scott Illingworth is a professor in the Graduate Acting Program at New York University's Tisch School of the Arts and a freelance director. He has taught, lectured, and directed at universities and schools across the United States and internationally. He is a member of the Stage Directors and Choreographers Society, a Guild Certified Feldenkrais Practitioner, and a Fulbright grant recipient. For more information about Scott, please visit www.physicalactioning.com.

Exercises for Embodied Actors

Tools for Physical Actioning

Scott Illingworth

NEW YORK AND LONDON

First published 2020
by Routledge
52 Vanderbilt Avenue, New York, NY 10017

and by Routledge
2 Park Square, Milton Park, Abingdon, Oxon OX14 4RN

Routledge is an imprint of the Taylor & Francis Group, an informa business

Scott Illingworth orcid: 0000-0003-4352-7868

Paperback cover image is a full brain tractogram by Sherbrooke Connectivity Imaging Lab (SCIL). Tractography uses specific MRI techniques and computer software to generate a model of nerve fiber pathways in the brain.

Library of Congress Cataloging-in-Publication Data
Names: Illingworth, Scott, author.
Title: Exercises for embodied actors : tools for physical actioning / Scott Illingworth.
Description: New York, NY : Routledge, 2020. |
Includes bibliographical references and index.
Identifiers: LCCN 2019049616 (print) | LCCN 2019049617 (ebook) |
ISBN 9780367433833 (hardback) | ISBN 9780367433840 (paperback) |
ISBN 9781003002819 (ebook)
Subjects: LCSH: Movement (Acting)
Classification: LCC PN2071.M6 I45 2020 (print) |
LCC PN2071.M6 (ebook) | DDC 792.02/8--dc23
LC record available at https://lccn.loc.gov/2019049616
LC ebook record available at https://lccn.loc.gov/2019049617

ISBN: 978-0-367-43383-3 (hbk)
ISBN: 978-0-367-43384-0 (pbk)
ISBN: 978-1-003-00281-9 (ebk)

Typeset in Sabon
by Taylor & Francis Books

For Jamie

Contents

Acknowledgements

My life is full of generous mentors and teachers. Allyn Sitjar, Kevin Kittle, and Lavinia Plonka were my earliest models of curiosity and rigor in theatre. The Theatre School at DePaul University in Chicago was a tremendous home for my advanced education and introduced me to Joe Slowik, Patrice Egleston, Phyllis Griffin, Betsy Hamilton, John Jenkins, Damon Kiely, and many others who all offered profound insight into acting as a craft and teaching as a profession. My time training to become a Guild Certified Feldenkrais Teacher® under Julie Casson Rubin and Paul Rubin transformed my view of the body and understanding of how the mind is part of the whole. Most of all I must thank Mark Wing-Davey and my colleagues and students in the Graduate Acting Program at New York University's Tisch School of the Arts. Mark's relentless curiosity coupled with his unwavering belief in me is the reason I have the space and time to continually develop this work.

I also owe a debt to a variety of people who helped in the process of writing this book. Specific thanks to Alexandra Templer for indispensable editing and research assistance, Mary Escobar for her graphic design work, Lisa Benavides-Nelson for her insightful feedback, and Mike May for his metaphor about train conducting.

Finally, I am mindful that none of those past or current opportunities would exist without my parents and sister, who never expressed a bit of doubt about my life in the theatre, or the unparalleled support of my wife, Jamie Wolfe, who is a partner in all things.

Introduction

Actors are makers. It is an important way to think about the work. Makers craft a thing over time until it is in a usable state for others. Making is a process and the process itself has everything to do with the quality and nature of the product. No matter how inspired the moment when you begin, it still takes time to make a thing. If I was building a table, for instance, just the vision in my mind of that table would not be the same as the finished table sitting before me. Many factors, from the materials I use to my own skills and experience, impact how close I will come to crafting the table I imagined. Makers are not magicians (which, in reality, is also a trade of making and refining illusions). Watching a performance can feel magical, though. Many of us get into the profession because of the experiences we had as audience members, but using a table is no closer to building a table than watching acting is to acting itself.

What I offer in this book are lots of tools to make your work – practical tools that will serve you in every version of the job I can imagine (from theatre, to screen, to the most last-minute audition, to virtual reality, to spaces we have yet to invent but where humanness needs representing). After using these tools, you can stand back and say "Wow, I made that table," instead of "Wow, I really love how I feel while making this table." There will be many moments of the joy of doing, but only because your focus is on the doing itself. I want you to experience the joy of doing it well and with purpose, not the joy of merely getting to do it. This is because, like the table, your work has a function beyond you. That function can be amusement or incitement to revolution or generating empathy or shock, or many other goals. Knowing the purpose of the thing you're making is an important part of making it.

In her iconic book, *Improvisation for the Theater*, Viola Spolin wrote "We learn through experience and experiencing, and no one teaches anyone anything."[1] In that spirit I am doing here the same thing I try to do in the classroom – provide exercises that are experiments for you to conduct. Experiments are questions built to test an idea. They don't presume an outcome and they leave space for the designer of the experiment to be proven wrong and try again. I crafted a process of inquiry you can use that happens, by the end, to engage you deeply physically, moment-to-moment, with whatever role you're playing and leaves you ready to experience the great joy of living in performance with others.

All the experiments spring from a vocabulary that both the experienced actor and novice share: action verbs. The vocabulary of actions is one you can use in any setting. Your director doesn't have to read this book for them to understand what you mean when you say that you want to "squeeze" your partner with a line of text or in an unspoken moment. They have their own embodied experience of that word, some of which likely overlaps with yours. Many actors find that when they start using this language with partners or in rehearsal others quickly pick up on it as a way to express their own impulses with greater specificity.

Part 1 through Part 3 of this book are all work you can do on your own. They are a personal set of tools. Part 1 introduces you to the core ideas about how you listen and respond in life and in your work through the lens of physical actioning. Part 2 guides you through a set of experiments to investigate yourself, your habits, and tendencies, and provides a chance to step back and be a more careful observer of this world that you both live in and hope to reflect. Part 3 applies that same curiosity and investigation to any role you might get to audition for or to play. You build an embodied role from first read through rehearsals. Part 4 offers experiments for the classroom and rehearsal hall where you and your collaborators explore these tools together on a production, scenes, or by honing specific skills. Learning acting works best in a group environment, but even if you're reading this alone, you'll discover a great deal you can apply in any rehearsal process.

If you're a teacher, you can also use the entire book to craft the arc of a course, as I do with my students. Many experiments throughout the book make good group classroom explorations to build experience and dexterity, embodying impulses. Importantly, all of the exercises employ the language of partner-focused action and offer ways to keep actors visibly physically engaged, even during early explorations with text.

Think of Part 5, the final section of this book, as a program note. It offers some context and a deeper dive by addressing two big questions that tend to come up about the connection to existing techniques and about the brain science referenced throughout. Linking new ideas about the mind to actor training has a long history. The title of the book calls you an *embodied* actor. Perhaps that seems obvious, but actor training historically put up walls between thought and embodiment or assumed they were connected but distinct ways of working. In the hundred-plus years since Stanislavsky built his experiments based on the emerging understanding of psychology, a whole revolution or two of sociology and neuroscience have exploded old theories about how humans work. Throughout this book you'll discover that this emerging knowledge about our brains offers exciting opportunities to explore new ways of working and provides tangible tools to move beyond the inside-out/outside-in myth of many acting methods. If we embrace it, neuroscience might even help us more reliably meet our core purpose of impacting audiences by illuminating how they experience performance.

The work outlined here is drawn largely from my classes with students in the Graduate Acting Program at New York University's Tisch School of the Arts. I'm acutely aware that the profession our graduates enter changes exponentially each year. Some of that change is a function of technology. Some is the aesthetics of the work. Some is a growing isolation in the process of getting jobs – often never even meeting people face-to-face. This new reality demands tools that actors can apply alone and with speed when necessary but that they can also crack open and explore deeply with others when the project and time permit. Importantly, these changes also include a recognition that the values of the profession long assumed that certain racial, gender, and other identities were "normal" or the "default." Many of these ideas are still imbedded in how we talk about the work and what we consider "good" or "truthful" acting. Addressing all of this requires a set of tools that is robust and flexible enough to apply to all of these different situations, beginning with the actor, and inherently valuing their unique point of view in collaborations. Like my students, I expect you will combine these tools with many others you've acquired along the way.

Dive into this work. Take your time. Some exercises ask you to do something for a week and then come back. You might repeat others a number of times before moving on. In an ideal world, we would have the opportunity to work together. The outside eye of a teacher providing feedback and pushing you is valuable. Most makers have an important period of apprenticeship where they can explore and fail and flail and grow with rigorous but encouraging feedback. Perhaps we'll meet someday. If not, I hope this book launches exciting explorations and frees you to become a more deliberately embodied and specific performer. At the end I have every confidence that you'll step back and see you made a table you didn't know you were capable of making.

Note

1 Spolin, Viola. *Improvisation for the Theatre*. Evanston, IL: Northwestern University Press, 1983, p. 3.

Part I

Foundations for Physical Actioning

1 Curiosity & Attention

Cobbling It Together

Part of what makes acting so difficult as a craft is the inability to stand back and see your own work. A cobbler (I know, not many around anymore, but stick with me) can finish a pair of shoes, take a look and compare them to other shoes, and, assuming they aren't particularly self-deluded, assess their skills compared to those of others. Even better, someone will wear those shoes who likely doesn't care much about the cobbler's feelings and happily tell them if the shoes don't fit or are poorly made or wear out too fast. The finished product of the cobbler's craft has a function that it must serve for the wearer/audience when complete. It does or does not serve that function well. Having the finished product separate from yourself allows a whole range of options for assessing your work.

Maybe cobbling isn't the best example. Perhaps it feels too utilitarian. "Shoes aren't art," you might argue, though I suspect Manolo Blahnik would disagree. Let's look at musicians instead. They have the tremendous benefit of musical notes. Notes vibrate at certain frequencies. That is measurable. You either hit the note or you don't. You are in time with the conductor or drummer or you are not. There is, of course, tremendous craft beyond that, but there are some satisfyingly objective assessments that a musician can use for calibration in performance and when listening back to a recording. Musicians occupy this interesting space where they both use their bodies as part of the work but still rely on tools outside themselves.

Maddeningly, acting is the craft of doing what we all do all the time – being human. The finished product looks (more or less, depending on the piece) like the rest of life. From that perspective, anyone can act. You're already human. You practice every day. All of your social interactions include some elements of performance. So why, then, are most people not very good actors even though they have all that experience under their belts? *Because most people don't really know what they're doing.*

We are living, not observing how we are living. We respond to the world, but don't often wonder about those responses. We operate from habit and pattern and most of life does not demand the kind of examination of intention or cause and effect that actors must apply to a performance. With all that practice at being human we are so habituated to our version of it we can't see how we *are* as humans. We don't recognize what we are doing moment-to-moment. We don't really see what is happening in others or between us and others. Often, we don't even know why we are responding the way we are to events and stimuli.

This is where training and practice must enter. To learn to act well you must make a study of humanness – your own and others. You must be a sociologist, psychologist, and neuroscientist. Acting is very real and utterly false. Part of that falseness is the repetition of events – knowing what is next. Sometimes it is the extremeness of the events – plays are full of traumatic and euphoric events that most actors never experience in real life. Your job is to become a ruthless observer of the true and a tireless practice-er of the skills that help the false appear to be true.

The appear part is important. Your experience in performing isn't useful if it isn't serving the story or reaching the audience. You have a responsibility to them. An actor can have a riveting experience for themselves that is uncommunicative or fails to produce empathy in the people watching. That is a frustrating complication of the work. My mentor and colleague, Mark Wing-Davey, argues that too often people seek to become actors because of how watching others act made them feel. Then they get on stage and try to experience the emotions that drew them to the work, failing to understand that the role of actor and the role of audience are different. Leave the response to the audience. Your job is the doing. Let your work produce empathy in others.

It is a daunting undertaking to study and replicate humanness. Perhaps you feel overwhelmed. Maybe you have a sudden interest in cobbling. Set those feelings aside for now. Turn your attention outside yourself and toward the doing. We'll begin with the first and most important trait of good actors. It is the mooring you should grab hold of if everything else floats away.

Curiosity

When people ask about the one characteristic I look for in an actor above all others I answer "Curiosity." I do not mean a passing, vague, selective, or self-serving curiosity. You must cultivate an insatiable curiosity for the world and others and information and contradicting points of view and everything that you encounter. Curiosity requires moving through the world with a quality of listening rather than speaking. It requires seeing the world in the form of a question rather than as confirmation for an answer you already possess. You must put yourself in places where the expected and the unexpected will cross your path and devour those experiences. The good news is that curiosity is something you can practice every minute of every day. Nobody needs to hire you in order for you to practice it and no job or task or conversation is without an opportunity to deploy it. The world is an endless stream of information about other people's humanity, social behavior, needs, fears, and how it all plays out in their bodies in space.

Curiosity is the foundation for your work and must exist within the work itself. What you can observe and unpack about the human interactions around you provides a rich catalogue of ideas, inspiration, and experiences when you confront a new script or a moment between characters that feels utterly unfamiliar to you. A state of curiosity can make learning more possible.[1] I always object when people suggest young actors cannot do good work because they haven't experienced enough in their lives yet. It assumes they've had an easy or privileged childhood and misunderstands our capacity to blend observation with imagination. If you are a voraciously curious observer of the world of any age, then you have the most important skill upon which to build the craft of an actor. No matter how old or experienced, a lack of curiosity means you pull characters as close to yourself as possible. You deny them their exciting distance from you. A failure of curiosity is the surest way to a failure of imagination.

How do you train curiosity? Leave space in your daily life to follow it. That means taking time to travel through your city or town in a way that few of us do in modern life. Your eyes are naturally curious, moving more frequently than you might imagine and drawn to the places with the most useful information.[2] Leave them free to discover objects and people in shared spaces instead of locking them onto a screen. There are many ways to be curious through technology, too, of course. Your work as an actor, however, requires practice cultivating curiosity in shared spaces without the mediation of technology. It means prioritizing your foundation as a theatre-maker over the comfort of disconnection from your immediate environment. The boredom you might feel creates space to listen to your own thoughts and where they take you. I'm pro-boredom. It's painful. An article in *The Atlantic* described one study where "two-thirds of men and a quarter of women would rather self-administer electric shocks than sit alone with their thoughts for 15 minutes."[3] On the flip side, another study suggested that the more time we have to exhaust the obvious answers, the more creative our responses become.[4] That's where you must reside.

Curiosity is also a foundation for empathy. Try this out. Find a wood floor and look at it. What can you see of the construction or the stain color or the flaws or cracks? How old is the floor? Who was the laborer that laid it down those years ago? Were they paid a living wage? What kind of wood is it? Where was that specific tree? Is that forest still there or was it all cut away for floors like these? Everything you see can inspire an internal "Why?" or "How?" Don't be satisfied with the simple answers. Almost every answer leads to new and more specific questions. Questions born from curiosity are a valuable way to see the way the labor and lives of others, often invisibly, intersect with you. Make it visible. There is still much we don't understand about how curiosity works, but evidence indicates that a high level of curiosity about an idea or topic helps develop memories about it and is associated with reward patterns in our brain.[5]

With that in mind, take a few moments now and do the following experiment about curiosity. This is the first of many experiments and exercises throughout the book. Whenever possible, I encourage you to really do them. Even if you can't do it right at this moment, come back to it before you read too much further. The experiments build on one another and often the text immediately following an experiment asks you to reflect on the experience in order to understand the ideas presented. You'll notice that for each instruction the first part is italicized and sometimes there is additional instruction that isn't. The italicized portions are the key step-by-step instructions for the experiment. In regular typeface are questions to ask yourself as you do that step or side-coaching that a teacher or director might offer as actors explore the instruction. It will become clear as you read that the instructions and additional text come from my experience teaching these experiments out loud with groups of students. I've tried to re-create that feeling for you here. If you're doing this work in a class or as part of a group, some of the experiments work best with one person taking the role of teacher or director and guiding the rest of the group by reading out loud. Most, however, can easily be done simply reading the instructions by yourself.

Rewarding Curiosity

1. *Select something you really enjoy doing but feel a bit guilty spending time on.* Keep it something simple like checking social media, or pampering yourself, or eating a cookie.
2. *Plan to do that activity right after this experiment.* Have the reward sitting nearby.
3. *In the location you're currently sitting, select an object or architectural element to investigate.*
4. *What gets your interest about it first?*
5. *Begin to ask questions about that item.*
6. *Start simply.* Perhaps you ask technical or more literal questions first. What color is it? What shape is it? Does it have marks or wear from use? What function does it serve in that location?
7. *Make sure to expand beyond the easily observable.* How long has it been here? How did it get here? What materials is it made of? Where do they come from? What people were involved in its creation? What are their lives and working conditions like?
8. *Be careful not to invent answers you cannot know.* If a question is answerable, that's great. Many are not answerable from where you are and with the information you currently have. The answers aren't the goal.
9. *Exhaust your ability to ask questions.* When you've run out, sit with the lack of questions and see what comes up. Try changing your position related to the item and let a literal new perspective help generate new questions.
10. *Once you've done this with the item for no less than 10 minutes, immediately allow yourself the reward you decided on before the experiment.*

Was it challenging to keep your focus on a single item for that long? Did your mind wander? To what? Did you feel irritable or frustrated or trapped by the task? Calmer? What feeling did the reward produce?

This isn't a one-time exercise. Active curiosity should become a daily practice. Perhaps, for you, the reward isn't necessary. Others may need help making curiosity a new habit in life and linking the act to patterns of reward in the brain. With time, you'll see how many moments each day offer the opportunity to practice and build this skill. Once you become a more habitually curious person, though, we must turn that into a tool you can use. The next step is taking that open and available curiosity and aiming it in a meaningful way.

Attention

Attention is where your curiosity points at a given moment in time. Curiosity is the state you must exist in. Attention is the conscious way you aim that state at a target. It is not merely focus or the performance of focus. Too often in rehearsal or performance actors lock their eyes on a partner's eyes – really trying to "see" the other. They often see nothing. They are performing an idea of attention. As Sartre wrote, "The attentive pupil who wishes to be attentive, his eyes riveted on the teacher, his ears open wide, so exhausted himself in playing the attentive role that he ends up by no longer hearing anything."[6]

Attention can mean looking into someone's eyes. More often, however, it means allowing different elements of their behavior, language, breath, clothing, walk, or whatever else might be observable to grab hold of your curiosity. Then, without locking onto that element needlessly, let the information you discover and your ongoing curiosity draw your attention to the next thing. You likely did this without even realizing in the first experiment. Your curiosity led somewhere specific. It provoked a discovery or question that brought you to the next thing. Your thoughts and ideas springboarded one to the next.

To understand how it operates in life, imagine this simple scenario. Let's say you are interviewing for a job. You might notice that the manager interviewing you is very expensively dressed. It makes you wonder if the company pays really well. You look around the room. The artwork is very nice, and the furniture looks expensive, too. Curious for further confirmation, your attention moves to a number of scuffs and chips on the lower corners of the desk. You look back to the interviewer and see that the very nice outfit is frayed at the seams. Your suspicion growing, you peer down and can see the bottoms of their shoes are quite worn. Is the business having a hard time? Was it once prosperous and now failing? Then your attention moves to sound. You haven't heard a phone ring anywhere and have been here more than an hour. This springboarding attention born from curiosity provides important information about the other person, the circumstances, and generates the next wave of curiosity. It builds on itself.

An actor must have deep and sustained curiosity about what they witness and how it is manifested in other people. Remember, when you feel lost or get locked on too tightly, take just one step back from attention and ask "What outside of myself grabs my voracious curiosity in this moment?" Then let your attention follow the flow of that curiosity. Let's take that earlier experiment in curiosity and build on it to include a conscious relationship to attention.

Springboarding Attention

1. *Find a public space where you know there will be quite a few people, but where you won't be expected to interact much.* Set a timer for at least 10 minutes before you begin so you don't need to think about the time.
2. *Let your eyes and ears start with a "soft focus."* This means that you're listening and seeing, but trying to take in the entire experience rather than one element or item.
3. *See what draws your attention first.* Is it a loud or predominant sound? Is it the sudden or large movement?
4. *Follow your curiosity to that item or event more directly.* Stay with it gently. While objects are okay to begin, ultimately let people be the main focus of this experiment. What specifically about the person or event gets your attention? Is that interesting enough to stay with? Maybe something else comes along that pulls your attention away by capturing your curiosity. It could be a different element of the same event, or maybe a new person or event steals it away.
5. *Keep going like this.* Continue to allow your attention to shift from item to item, person to person, event to event. Stay soft enough to be open to new stimuli while maintaining curiosity about whatever your attention is aimed at in the moment.
6. *When the timer goes off, take a few minutes and make notes about all of the things that had your attention in the last ten minutes.*

How many items can you remember? Why did those items stick when you forgot others as soon as they were gone? Which ones held your attention the longest? Is there any similarity between them? In a way, I'm asking you to do an anti-meditation. Instead of letting things cross through your mind and let them go, I want you to grab on to them and ride them to see where they take you next.

Like many things, when you first do this it can feel quite technical. If you learned to drive a car you know that early on it takes tremendous conscious attention to do all the elements of that task. Where are my hands? Eyes on the road. Check the speed. Remember to look at each of the mirrors every few seconds. Eventually you carry on conversations and listen to music and drive somewhere and remember almost nothing about the drive. But then there are those moments when you get lost. That's when "Everyone needs to stop talking and can someone please turn down that stupid music?!" The demand on your attention rises and all of the other things that might take it must disappear.

The same thing is true of attention on partners in your work. With practice, curiosity about partner leads the flow of attention moment-to-moment. Attention helps you focus on the important events and assess how they might impact you.[7] It may seem silly to talk about human interaction in such a basic way, but the very fact that this constant stream of information and attention happens so quickly and with so little awareness makes it worth stepping back to investigate. Awareness helps you build choices. *You can't do something in a new way if you don't know how you do it.*

Now you've got two of the three key elements you need: curiosity and attention. Of course, you could let your curiosity point your attention all day long and still remain a passive observer of events. Your work isn't about watching events, it is about engaging in them. The final element is the focus of this book. It's what happens when you're not just standing off to the side observing. It's what happens when the target of your attention does something to you, wants something from you – when you're in the event together. In life and in performance your curiosity that leads your attention is not the point. It is in service of sparking your response. This detailed and ever-evolving stimulus you're now attuned to becomes the source of the one thing you control as an actor.

Notes

1 Gruber, Matthias J., Bernard D. Gelman, and Charan Ranganath. "States of Curiosity Modulate Hippocampus-Dependent Learning via the Dopaminergic Circuit." *Neuron* 84, no. 2 (2014): 486–498. DOI: 10.1016/j.neuron.2014.08.060.
2 Madary, Michael. "Visual Experience." In *The Routledge Handbook of Embodied Cognition*, edited by Lawrence Shapiro, 1st ed. New York: Routledge, 2014, pp. 263–271.
3 Stewart, Jude. "Boredom Is Good for You." *The Atlantic*. Atlantic Media Company. June 7, 2018. Accessed March 20, 2019. https://www.theatlantic.com/magazine/archive/2017/06/make-time-for-boredom/524514/#10.
4 Schubert, Daniel S.P. "Boredom as an Antagonist of Creativity." *The Journal of Creative Behavior* 11, no. 4 (1977).
5 Gruber, Matthias J., Bernard D. Gelman, and Charan Ranganath. "States of Curiosity Modulate Hippocampus-Dependent Learning via the Dopaminergic Circuit." *Neuron* 84, no. 2 (2014): 486–498. DOI: 10.1016/j.neuron.2014.08.060.
6 Sartre, Jean-Paul. *Being and Nothingness: An Essay on Phenomenological Ontology*. Oxford: Taylor & Francis, 2013, pp. 83–84. ProQuest Ebook Central. Accessed March 23, 2019.
7 Blair, Rhonda. *The Actor, Image and Action: Acting and Cognitive Neuroscience*. New York: Routledge, 2008, p. 61.

2 Action

What Is an Action?

I mean a few seemingly different things when I use the word "action." This is on purpose because, as you'll discover, they are all connected in our bodies, though we often speak about them as distinct. The most common use of action could roughly be described as *behavior* or *activity* – something you consciously, physically do. A lot of what you observed in the previous chapter's experiments probably falls into this category. Someone looked at their phone or picked up a scrap of paper or took a picture. In other settings, you might hear teachers or directors refer to actions in relation to text or speech. You "play an action" on a line or word. It describes something about how you express those words and the desired impact of the words on your partner. This is a great tool for script analysis and certainly a part of what I mean when I speak about actions, but it is also not the full picture.

The way I use the term "action" in this book is a more holistic or combined view of these forms. *Actions are the totality of the thought, speech, and all other physical manifestations for a particular impulse.* This includes gestures, behavior, walking, breath, muscle contractions, and much more. Thoughts are physical, too. They happen in your brain (a part of your body) and are measurable chemical and electrical events. Some studies indicate that simply thinking about an activity or hearing others describe it sparks some of the same parts of the brain as actually doing that activity.[1] In training actors, teachers often use phrases like "Get out of your head," or "Don't think, do." Not only is the metaphor unhelpful, it just isn't accurate. What happens in your body shapes your perception and thoughts just as your thoughts begin to activate the rest of your body.

You Read Minds

You may have noticed in the last experiment that, as you followed your curiosity and placed your attention on various people, a certain set of beliefs or guesses about that person emerged. How did you land on those ideas? You started gathering information that told you something about their life.

You watched their behavior. A countless series of associations and previous experiences and biases and knowledge are part of these instant assessments. You couldn't possibly understand or be conscious of all of them. You can't put your attention on everything at once. How do you determine what information to gather in your observations and come to your opinion of the person and what they are doing? You watch their visible actions. A voraciously curious actor isn't really a mind reader, but an *action* reader. The curiosity that draws your attention in life and performance leads you to observe what the other person is doing. The "doing" can be activities, a face they make, a shift in eye contact, words, gestures, sounds, a sigh, or whatever combination of these and more might be observable to someone else. These are the visible manifestations of their action. You take in all of this information moment-to-moment and try to put together what it all means.

People Are Icebergs

Why do you do this? Why, when you simply see a person picking up a stick, do you look for a dog nearby or figure out some other cause? Why isn't simply seeing the event sufficient? Because we are built to assess intent. Our brains constantly work to understand the "why."[2] We recognize that we only see the outward expression of some larger process – the tip of the intention iceberg. We know, deep down, that nothing happens without some need or want behind it. As you turn your attention to people and events in the world, you build a story of cause and effect. It is simple to see some cause and effect events. If a branch falls and almost hits a person and they jump to the side, you have enough experience to sort out that they jumped to avoid being injured by the branch. Now imagine you see a person from a block away on the sidewalk. As they get to the crosswalk, they jump over something and continue on their way. Even if you can't see what they jumped over, you understand that there is some object to avoid. If it had been raining all day you might feel certain it was a puddle. The parts of their actions you can see help you answer the constant "why" in your mind.

Until now you have focused on observing others in situations that didn't involve you. Your work as an actor, of course, happens mostly in situations where actions are aimed *at* you. Often those actions are much more subtle than leaping out of the way of falling limbs. It is important to recognize that actions underpin even small social behaviors. Think of someone you know who smiles all the time – it comes easily and readily when they see someone they know, at waitstaff in a restaurant, upon meeting new people, and in every photograph. If they enter the room and smile at you, your curiosity draws your attention to that smile – that physical manifestation of their action. You instantly develop a point of view about why they are smiling. Are they happy to see you? Does it seem a little forced today? Are they are hiding something? Does the smile match their tone of voice and the rest of their behavior?

We do not receive actions neutrally. We have a point of view based on our previous interactions with this person, the many interactions with other people in our life, and events we witness or even hear about involving people with similar traits as this person. All of those combine in that instant while we try to understand not just what the person is doing, but *why* they are doing it. It happens faster than we can even fathom. A 2014 study by researchers at the Massachusetts Institute of Technology (MIT) demonstrated that humans can recognize and identify an image flashing by in as little as 13 milliseconds.[3] Imagine how much information you process at the much slower speed of a smile or a handshake or a kiss.

Actually, You're a Terrible Mind Reader

Of course, we are not only action readers, we are also mis-readers. In fact, that might be far more common than not. There are simple errors. I, for instance, make jokes in a very dry manner. Some people pick up on the small signals that I'm kidding right away. Others have a hard time knowing until they've spent more time with me. Frequently, several weeks into the semester, a student will say they have just realized that I'm being funny. That's a fairly innocuous example. In many circumstances misreading actions produces grave results. Think of all of the instances of police officers shooting unarmed people holding a cell phone or candy bar or nothing at all. For a range of reasons, personal and systemic, the police misread their actions and tragedy results. This is a painful example of how much can be at stake as we read and misread the actions of others. It is also the heart of theatrical conflict. Plays and films are full of moments when one character misreads the action of another, responds to what they *thought* was happening, and heightens or creates a conflict.

The Action Loop

The three key elements: curiosity, attention, and action combine to create a cycle we engage in continuously as people and the core loop you must live inside of as an actor. You exist in a state of curiosity that leads your attention, and your attention brings you to witness an action by your partner, which you read. And then what? *You take action in response.* You are not merely a sponge that absorbs all of the actions that come your way. The actions of your partner spur you to respond. This completes the cycle because, inevitably, you become curious about how the other person responds to what you just did. That is the full loop.

For a simple example, imagine buying a coffee. You order (action), you watch how they respond (curiosity and attention), they tell you the price (their action), you hand them exact change (action), you watch them count it (curiosity and attention), they give you a coffee (their action). Obviously, depending on the details, there could be many more actions within that exchange than the simple

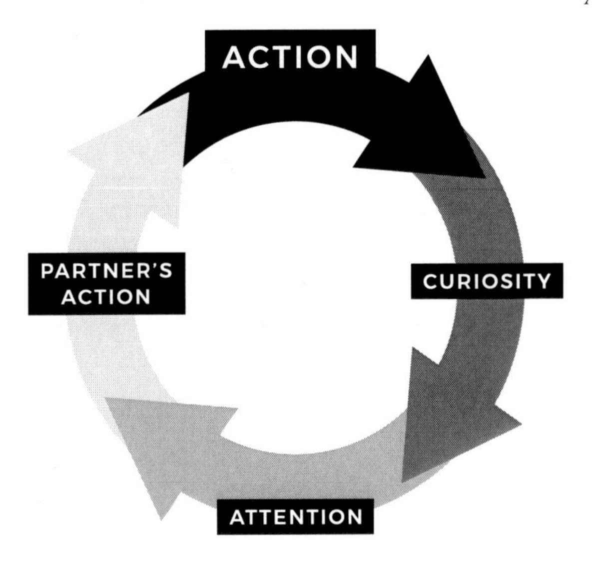

The Action Loop

ones I describe. In addition, you can probably see how many ways it might go awry. Imagine that the cashier takes your money and then walks away without giving you your coffee. What action does that inspire in you? What if they skip taking your money and just give you coffee for free? What is your action in response? We could also look at the entire exchange from the point of view of the cashier. They have their own series of actions, which may start when you walk in the door rather than when you order the coffee.

Not only is this process happening at the social interaction level, your brain engages in a similar loop as it tries to predict what is going to happen next and models the world around you. You use your body in action to develop those predictions. As Andy Clark writes in *Surfing Uncertainty*, "Perception and action are thus locked in a kind of endless circular embrace."[4] This simple loop of human interaction is also the core of the audience experience. Person A did this. What will Person B do in response? Ah, they did that. What does Person A think of that response? What will they do? And off we go.

In some ways we could end the book here. You cultivate voracious curiosity. You turn that into attention aimed at a partner. You witness and read their action. Finally, you take action in response, and return to curiosity and attention to see how they respond. That is the core of every alive moment of performance for both actors and their audience. This loop keeps your focus off of your own experience and requires that your partner remains the source of and inspiration for everything you do moment-to-moment. Of course, if you're holding this book, based on the pages in your right hand versus your left, you can tell that we've barely begun. So much complicates your ability to do the work simply. You have a lifetime of habits and deep grooves of behavior that stand in the way. There are obligations to the storytelling. There are technical demands. You have ideas about delivering a performance. You have those ways you want acting to feel. We'll do many experiments to unpack all of that and build tools to reliably live in this loop of actions with ease and confidence and without excess effort.

Notes

1 Garner, Stanton B. Jr. *Kinesthetic Spectatorship in the Theatre: Phenomenology, Cognition, Movement*, Cognitive Studies in Literature and Performance. Cham, Switzerland: Palgrave Macmillan, 2018, p. 154. DOI: 10.1007/978-3-319-91794-8.
2 Gallese, Vittorio. "The 'Shared Manifold' Hypothesis: From Neurons to Empathy." *Journal of Consciousness Studies* 8, nos. 5–7 (2001): 33–50.
3 Potter, Mary C., Brad Wyble, Carl Erick Hagmann, and Emily S. McCourt. "Detecting Meaning in RSVP at 13 ms per Picture." *Attention, Perception, and Psychophysics* 76, no. 2 (2014): 270–279. DOI: 10.3758/s13414–013–0605-z.
4 Clark, Andy. *Surfing Uncertainty: Prediction, Action, and the Embodied Mind.* Oxford: Oxford University Press, 2015, pp. 6–7. DOI: 10.1093/acprof:oso/9780190217013.001.0001.

3 Translating Physical Language

Why *Physical* Actions?

If we want to find a unified way to describe all of the elements that come together in a particular moment to form an action, we need a vocabulary and set of tools that embraces an integrated self. Many actors and acting teachers still talk about working "from the inside out" or "from the outside in." This comes from old ideas about the relationship between psychology and the body. When that vocabulary first developed, it reflected how people believed emotions and behaviors developed into physical sensations. In reality it is far more complex and less like a simple two-way street. Thoughts have physical manifestations and external physical acts alter the brain, including our thoughts and emotional responses. It is a single integrated system.[1] You can no more separate your mind from the rest of your body than you can pull the color red out of a brick. Even when people refer to actions or physical actions many times, they still maintain other categories that preserve this separation. For instance, Stanislavsky spoke about "physical action" but he also spoke about "verbal action," differentiating those impulses.[2]

Rather than imagining thoughts, speech, and movement as distinct elements to consider, I like the term "physical actions" to capture all of these at once. It reminds you that every impulse is embodied. As we discussed in the previous chapter, an impulse can include some combination of speaking a line of text, other sound and breath, conscious or unconscious facial expressions, movement in the rest of the body (like walking, gesturing, activities, dropping to the floor), and internal sensations that might not be visible to an audience or partner (like a change in heart rate or the hair standing up on the back of your neck). That's a lot. It should include everything. Moment-to-moment interactions are rich with complexity. If we are going to try to capture all that humanness within our work, we need a way to express these complex responses quickly and simply without losing how specific they are.

Give It a Name

You could probably write pages attempting to describe exactly what goes on in you when a roommate or sibling says "Are you done in the bathroom?"

For your work as an actor, it is important to develop a simple and clear vocabulary you can use for yourself and in conversation with collaborators. It shouldn't require everyone to take the same acting class or read the same books or even believe in the same ways of doing the work. You need something that efficiently describes all the ways your and your partner's actions might manifest themselves. You'll see, as we progress, it is also essential that this vocabulary embraces the embodied nature of your thoughts right from the start to defeat that imaginary thought/voice/body divide that so many actors struggle to overcome.

The simplest way to talk about actions is to use the seemingly limitless number of verbs available to describe what people do to one another. For our purposes, we'll describe physical actions by using *transitive verbs*. What is a transitive verb? Basically, it means in order to work, the verb has to have some target to land on. Most of the time if you can put it between "I" and "you" (as in "I worship you") or you can put it between two names (as in "Niko worships Jordan"), that usually does the trick. Why pick one way or the other? Sometimes the "s" at the end of a verb helps you confirm it is a transitive verb. It is important that you don't need to add extra words to make it work. For instance, "I work with you" isn't great. Or even "Niko pleads with Jordan" isn't as useful.

This is because of how we hope to use the verbs. Think about "working with" someone. In some respects, what they do or don't do in response to your action is irrelevant. I think we've all had co-workers that did nothing. Sure, you worked *with* them, but really you worked on your own alongside them. The extra words make it harder to know if your action succeeded quickly. It's better to use "I help you" or "I aid you." You can watch you to see if your partner seems helped or aided. Did you succeed? Is it working a little bit? What signs tell you the answer? It keeps your curiosity and attention on your partner for the answer, rather than a focus on you and your "working with." *The key to physical actions is that the test of their success or failure is in your partner.*

The same is true for that second example. I suppose you can watch your partner to see if they seem "pleaded with." The problem is how easily it allows you to simply perform an *idea* of "pleading" without a test on your partner. "Niko begs Jordan" is better. It is more immediate, clearer, and the verb lands directly on the partner. The simpler the construction the easier it is to play the physical action, see it land (through curiosity and attention), and assess your success or failure based on the physical action they do in response. The great thing about actions is that they're personal. The verbs that spark you might be different from the ones that spark someone else. As you observe the physical actions of others, you'll come up with words that reveal not just what they are doing (and how clearly) but also a great deal about your own point of view.

Importantly, these verbs are also inherently embodied. You can't think of "skewers" without getting an image of that thin piece of metal sliding through veggies or meat. Simply hearing or reading verbs produces a response in your mind related to motion and/or events.[3] Using verbs in this way begins the imaginative process of doing them. Now think about what it might mean to "skewer" a person. Maybe you thought about literally skewering someone – a horrific image. Maybe it generated thoughts of saying something cruel to a person that pierced them in some way. Whatever came to mind, imagining that verb generated an embodied idea of the impact you wanted your words and behavior to have and ignited some of the same pathways in your mind as actually doing it.[4] This work involves coming up with many delicious words to describe what you and your partners are playing. I highly recommend *Actions: The Actor's Thesaurus* by Marina Caldarone and Maggie Lloyd-Williams as an indispensable companion for these experiments. In case you don't have it, I have provided my own short list of action verbs at the back of this book to get you started.

Let's begin by working on a simple six-event scene. If you can, do this in a group. Don't let anyone but the two players see the script. Once it is performed, give everyone else a copy of the script and have them write the physical actions they think they saw played. It often helps to have the actors play the scene a few times in a row, attempting to do it in the same way as far as possible. If you're not working in a group, simply read the scene below and imagine the events a few times before writing down the physical actions.

Naming Actions in Behavior

Below is the observable behavior and speech of two people interacting. Go ahead and name the physical actions you think they play. Select just one transitive verb from the list in the back of the book for each of their "lines" of text/behaviors.

A: Enters a coffee shop, sees B sitting at a table by the window, smiles and waves.
B: Waves back at A, half gets up, and gestures to the open chair across from them.
A: Approaches the table and says "Here we are."
B: Chuckles.
A: Doesn't sit.
B: Sits back down.

Here are two (of many) possible versions of the physical actions for this scene:

Version 1:

A: Alerts B
B: Encourages A

A: Warms B
B: Joins A
A: Questions B
B: Invites A

Version 2:

A: Tolerates B
B: Allows A
A: Tests B
B: Humors A
A: Confronts B
B: Dismisses A

What You Do Reveals the Why

You may have come up with some physical actions that were quite similar and others that were entirely different from my examples. If you watched people play the scene, the physical actions you read from their behavior and speech likely told a complex story about relationship and history. You filled in the blanks of "why" by working to make sense of their physical actions. If you were reading the scene from the book, to sort out the potential actions you probably imagined specific people, invented characters and relationships, and may have assigned A and B with genders or other group identities. Each physical action grows from some set of information (known or invented) about the relationship between these two people and the history that they share. In addition, each physical action is a response to the one before it. Together, they create a Score of Physical Actions that not only tells us the story of the scene, but also reveals the character of each person in it.

This short script just contains "what" happens. It tells you what the people say and some of what they do during the scene. A more complicated script sometimes reveals why someone is doing the things they do through dialogue or events in the plot. In the scene above, perhaps they are strangers and one person is here to sell the other their used car. Or maybe they are estranged siblings. Whatever the circumstance, you as an actor must sort out the "how." *Physical actions are the roadmap of how*. They help you determine how this person does and says the things in the script to achieve their goal by giving you something to *do*.

How Is the Result, Not the Choice

A key reason you select physical actions is to focus on the doing itself. Without something specific to do, you risk getting caught up in an empty version of what the moment should look like. You lock into a pattern of inflection, or a gesture – all the outward expressions of physical actions without the action itself. You generate the external shape of your performance and become focused on the audience rather than your partner. I see this in auditions all the time. An actor comes in and provides a performance that is empty and technical. Although I can see from the gestures and inflections that there was a moment in their preparation when this version of the piece was quite exciting, what they've done is maintained the shape of that performance, but all that is left is the shell. Everything that was driving those choices is gone and so the work seems hollow. Too often actors lean on some outward expression they think fits the moment or that a director or coach affirmed in an exciting discovery during rehearsal. They understand *how* they did it, so they simply repeat that external form. They don't understand the physical action that generated and unified that outward expression. The solution to this is to build a vocabulary of physical actions. The core of your performance must be the actions you play constructed on a knowledge of the piece and clarity about your desired response from your partner. If you do that, the expression of that action can shift or solidify or grow or ebb and the life of the performance will remain. *Let "how" be the by-product of the physical action.*

Actions Must Add Information

It is important to guard against selecting physical actions that merely describe the event or restate the line of text. In our sample scene, for instance, when A comes in and waves, the action "A greets B" isn't very useful. The same would be true if A came in and had the line "Hello." "Greets" doesn't add anything to our understanding of the moment. Even if you believe that it isn't a profoundly dramatic moment, something like "A signals B" leaves more room to play. Certainly, something like "A humors B" or "A warns B" set up lots of interesting paths for a scene. Even without any more context than provided in this simple scene, you can ensure that your action demands something specific from you and from your partner in response. Your physical action should be more than an errand, so that you really have to look for the response of your partner. How did you impact them? Even the simplest behavior has a "why" behind it. Remember that the brains of your partners and your audience are always looking to understand the reason behind what is visible. Actions should help the audience answer the question "Why?" not merely repeat the same information they'd get by reading the script.

Small Changes in Action Transform a Story

Try the experiment in another way now. If you have a group, have two new actors play the scene just like the last pair, only change the last two physical actions in the scene. Don't change the script, just the verb/action they play. How does that one change alter everyone's experience watching the scene? What new or different story does it tell? What expectations does it establish for what might come next? If you're reading this book on your own, simply try a few experiments with your selections. What if you change just the final physical action? What if you keep A's actions the same and change all of B's? What new story does it tell us as an audience? What questions does it open up about the relationship?

If you're doing this in front of a group, I discourage you from trying to make the scene too "interesting." If you find yourself picking a bunch of actions that you believe will make the audience in the room laugh, I suspect you're not really playing physical actions on your partner. You're probably playing the action "amuses" and have made the audience or the teacher your partner – leaving your actual scene partner to fend for themselves. Make the audience lean in to understand the event. Physical actions must be played on your partner, not just for the benefit of the audience. Don't act *alongside* your partner, act *with* them.

This tiny theatrical event (someone meeting someone else at a coffee shop with only three words spoken in the whole interaction) is an entire play. As viewers, we automatically work to fill in the gaps. At first you may find the idea of selecting an action for every line or event incredibly burdensome or limiting. You might even notice that the actors seem less spontaneous in their performance of this tiny scene. I don't advocate that you select a physical action for every moment in advance. As the book proceeds, you'll get a better sense of how and when this tool applies in a moment-by-moment way. For now, the important discovery is how significant any single choice is for storytelling. Changing a single action can fundamentally alter what comes next, the way your partner must respond, and directly impacts the experience of the audience and their expectations.

You Built the Foundation

As you now know, your work as an actor must begin with a voracious curiosity about the world. It is something you can cultivate and practice in and out of the work itself. That practiced curiosity then becomes aimed at your partner in the form of attention. This attention allows you to witness and attempt to read their physical action, which, in turn, inspires your physical action in response. This loop becomes the core moment-to-moment cycle of your life in performance. To help you describe all of the ways physical actions manifest in the simplest unified way you use transitive verbs. These verbs are clear to everyone you work with, even though they possess profoundly personal subjective meanings. They also remind you that thought itself is action and that ideas and impulses are embodied.

Now that you have these three core ideas in mind, we'll explore how they operate within your work. We'll begin the same way I do with my students in class, by examining the physical actions you play in everyday life. A careful examination of yourself paves the way for nuanced and specific choices when you turn your attention to imaginary circumstances.

Notes

1 Kemp, Rick. *Embodied Acting: What Neuroscience Tells Us about Performance.* New York: Routledge, 2012, p. 100.
2 Benedetti, Jean. *Stanislavski and the Actor.* New York: Routledge, 1998, p. 87.
3 Bedny, Marina, Alfonso Caramazza, Emily Grossman, Alvaro Pascual-Leone, and Rebecca Saxe. "Concepts Are More than Percepts: The Case of Action Verbs." *The Journal of Neuroscience* 28, no. 44 (2008): 11347–11353. DOI: 10.1523/JNEUROSCI.3039-08.2008.
4 Kemp, Rick. *Embodied Acting: What Neuroscience Tells Us about Performance.* New York: Routledge, 2012, p. 110.

Part II

Seeing Yourself

4 The Physical Action Journal

Living Inside the Experiment

Actors aspire to be truth-tellers, but most people do not see themselves, others, or their interactions with much clarity. It's hard to recreate a reality you don't actually see. I call the exercises in this book experiments because it serves as a valuable way to explore the work. We test a theory or a question in a controlled way and observe the results. There isn't necessarily a right or wrong result. The primary purpose is to get new information from the experiment and we expect to repeat it many times. A good artistic process mirrors a good scientific process in many ways. They both demand curiosity and they both require a willingness to try things out, be proven wrong, and come back with a more refined or different question. Sometimes they both (painfully) demand throwing out weeks or months or years of work and trusting that what was discovered through failure will launch a better exploration next time. The challenge to this comparison is, of course, that a proper scientific experiment doesn't have the scientist as the subject of the experiment. A hallmark of that process is objectivity in analyzing the results. Acting requires the constant use of yourself in an attempt to recreate a version of life itself with all its contaminating variables. That causes some messy complications to our experiments that we must confront.

We know that one problem with observing ourselves is how little awareness we have of what we do moment-to-moment in interactions with others. Even when we observe our responses with clarity, it may still leave *why* we did what we did a mystery. We are, so often, on a kind of autopilot. We react to people or environments habitually rather than seeing the unique features of an interaction or stepping back to probe why an exchange provoked the response it did from us and others. Stanislavsky thought "some 90% of normal behavior was automatic."[1] Neuroscience continues to reveal new information about the "automatic" part. In some respects, we may not even interact with the world that actually exists. Our brains are constantly engaged in predictive processing, producing a kind of simulation of the world based on past experiences and sensory input.[2] Just imagine what this might mean if you repeat the same moment dozens or hundreds of times with a scene partner over weeks or months of rehearsal and performance. How much of what they do are you even really experiencing anymore? How much of your own response is truly responsive? No wonder it is so hard to "stay in the moment" as directors and teachers often instruct. Stanislavsky's estimation of how much we consciously control is probably too generous. We all have experiences of being certain of something and then being suddenly confronted with a different reality. I have a vivid memory of being in high school and seeing a dear friend from across the room. I ran up and gave him a hug only to experience his resistance. I looked again, and saw the *actual* face of the person – a total stranger. My brain predicted my friend. I *saw* him. And I was embarrassingly wrong.

So, what can we do to overcome this? These predictions our brain makes are based on previous experience and tested, moment-to-moment, against the sensory input from our physical interaction with the world.[3] As an actor, finding ways to stay in the moment means finding ways to put your conscious attention on things that might otherwise slip past because you know so much about what is supposed to happen next. It means focusing on that sensory input with a different kind of attention than you usually deploy. You must live in the experiment and also try see it with the clarity of the observer.

That's why, rather than beginning with the imaginary circumstances of a play (where you likely have an entire set of values and ideas about performance and good acting lurking), we'll begin with your very real life. The experiments in Part 2 stoke curiosity about your interactions with others and move elements of those experiences from a background to a foreground process. You'll develop comfort with living inside the questions "why," and "how," far more than normal and develop a vocabulary to describe what you discover. A major influence on my directing and teaching is the work of Moshe Feldenkrais, developer of The Feldenkrais Method. He is often roughly quoted as saying "If you know what you're doing, you can do what you want." Let's investigate what you're doing by taking your curiosity and aiming it at the deceptively simple question "Why did I do that?"

Field Notes about You

A number of years ago I was working with some students discussing the physical actions they most frequently play in day-to-day life and one suggested that they all try to keep a Physical Action Journal for a week. It was a great suggestion and the results were fascinating. Student after student came in to the next class horrified by their list. They felt shocked by how manipulative they seemed through this lens. Some spoke about interactions with roommates where they asked seemingly innocuous questions like "What time are you getting up tomorrow?" only to realize under scrutiny that their physical action was something like "chastises" or "tests." Others recalled taking a hard line with a family member in an argument but discovering that their physical action was really something closer to "allows" or "pardons," suddenly revealing why they keep having the same fight over and over. It appeared as if their whole existence was tricking people into doing things they wanted done or completely caving in to the impulses of others.

I have a more generous view of their discoveries. In nearly every interaction with others each day we simply try to get our wants and needs met. Without scrutiny, we casually ascribe innocuous and low-stakes motives to our behavior. Our attention is generally on what we are doing and not our body as we do it.[4] Once we turn our conscious attention to these little interactions and begin to name what we are doing it reveals a world of needs and expectations we hide from ourselves. Stanislavsky knew that even thought itself had motivation.[5]

A simple example might happen at the checkout counter of a grocery store. If you walk up to the cashier and they are turned away from you distractedly talking to their neighboring worker you might say "Hi, there." Perhaps your physical action here is "I alert them" or "I prompt them." What if you don't say anything? Maybe you just put the stuff from your basket on the counter a little more loudly than necessary, or cough, or make your cart slide against the rack of candy and clang a little. "Alert" and "prompt" could also apply to those entirely non-verbal versions. *Any given physical action has a myriad of ways to be played.* Are you manipulative or mean for needing to check out and carry on with your day? Probably not. Of course, if you did any of the same verbal or non-verbal behaviors and your physical action was "I scold them" or "I abuse them" that would communicate a whole different set of things to the cashier about you. It might also mean that it takes you even longer to leave the store depending on the actions they play in response. Just as you can play physical actions in many different ways, *the same language or behavior can be the manifestation of many different physical actions.*

Here's a slightly more complex example with a roommate or family member. Imagine that they eat a treat you're saving for a special moment. You open the refrigerator, see that it is gone, walk over to them and say "Was it good?" Or maybe you simply buy a replacement treat and write your name on the package of the new one. Or perhaps you sit down next to them, pick up the empty wrapper, and jokingly look at it with melodramatic sadness and longing. All of these versions might be the physical action "I train them" or "I correct them." Some versions are more direct ways to play those actions and some are more passive, but there are potentially hundreds of ways to accomplish those physical actions in that circumstance. If this is the tenth time they have eaten your food, perhaps your action is more like "I slap them." On the other hand, given how long you've allowed it to go on without addressing it directly, maybe an honest look at yourself reveals that your physical action is "I permit them."

Witnessing even this brief interaction between you and a roommate or family member reveals a lot about that relationship and your character, not just by *what* you do and say but *how* you do and say it. The combination of those two offer an observer hints about the physical action you're playing. You play the physical action. Your partner (and anyone watching) receives the "what" and "how," and then tries to read your underlying action from that. It is easy to overlook these small exchanges in life and actors often ignore them while working on a role in favor of major emotional events. We know, however, that all those brains in the audience take in verbal and non-verbal information every second. They experience your character even when you don't intend them to read your behavior. You must play clear and specific physical actions that communicate character in each moment.

Notice, none of the actions I suggested for this example are "I question them" or similar. As you think about physical actions from your life, be careful not to simply restate the event itself or find a verb that just describes what you did and said. As we discussed in Chapter 3, the action illuminates the "why" and generates the "how." The easy part is recognizing the "what." In a way, the Physical Action Journal asks you to reverse engineer the way you work with a script. There, you get the "what" and the "why" provided by the author and must fill in the "how" as part of your work. In these next experiments looking at yourself, probe for a deeper understanding of why you do what you do in given moments by growing your ability to observe the behavior itself.

What students respond to in horror after keeping a Physical Action Journal is the idea that they are choosing these actions in life, moment-to-moment. I believe they are. We're usually not putting a name to it and the selection happens so quickly, in the background, and based on so many previous experiences, that we rarely recognize the moment of choice. Even if most of the physical actions we play in life are not extreme or dramatic, we still have impulses and those impulses become embodied through speech and non-verbal physicalization.

It is difficult to become an observer of yourself. You must be willing to both acknowledge the need you're trying to have met in a moment and also be generous enough not to assume every impulse you have is a grand scheme of world domination. To start this process, you'll take some time to observe yourself through this lens and practice with the vocabulary of physical actions. Don't worry about how quickly you can come up with the words or whether or not you picked the right one. This is likely your first attempt to apply a unified vocabulary to name the way you embody thought impulses. It will take time. Lead with curiosity rather than judgement or certainty. It can also be great fun to stand slightly apart from your own life and try to observe the way you interact with others. Here is your first task.

Physical Action Journal

1. *Keep a notebook for a week.* Carry the journal with you wherever you go so you can make observations in the moment.
2. *Each day, write 8–10 physical actions you play.* Remember, not every example needs to include a time you speak. Some actions are played without speech and all actions include your entire body in various ways.

3 *Keep two columns for each moment you observe.* In one column write the (transitive) action verb. Use the other column to make a note describing the verbal and non-verbal ways you played those actions.
4 *If you can't come up with the verb, just write in the description column.* This is especially good for moments when you're very interested in what motivated your behavior but cannot quite put a name to it yet.
5 *Phrase them as "YOUR NAME [transitive verb] THEIR NAME"* to get a little observational distance (for instance, "Peyton schools Shawn"). For now, make sure there is an "s" at the end of the verb.
6 *Try to note moments with people from different parts of your life*: personal and professional, intimate and stranger, and higher and lower status as you perceive it.
7 *Make certain that some physical actions come from experiences where you walk into the conversation knowing you are going to observe your actions and others where you catch yourself playing the actions.*
8 *You can use the short glossary in the back of this book*, but feel free to think of action verbs yourself. Feel free to use words the way you use them. Slang and/or alternate uses are totally fine. If you've struggled to keep a journal during a given day, try looking at the list in the back of the book that evening to see if any verbs jog a memory from the day.
9 *Don't tell people you're doing this with them*, as it changes you and them.

At the end of the week you'll have nearly 70 physical actions you played in a variety of circumstances with a range of people. That's a great start. Now, sit with the list and read it through. Are there physical actions that you repeated most days or in a variety of situations? Are there others that you could group together as synonyms or in a kind of family of physical actions? Often when people do this experiment, they begin to recognize patterns in certain relationships or even across their entire lives. You probably have a variety of templates for responding to certain types of situations. When a new, but similar, situation arises, you readily engage that habitual response, often without even realizing it.[6]

You might also recognize that certain physical actions assume a certain status in your mind. For instance, if your list is full of verbs like "halts," "disses," "soothes," and "welcomes," perhaps you operate from a fairly high-status place. Even a potentially generous action like "soothes" assumes the ability to provide some comfort and reassurance to others. If, however, your Physical Action Journal is full of words like "elevates," "admires," "permits," and "dodges," maybe you're operating from a lower-status place. Higher- or lower status versions of all of these exist, of course. Use this first look at the Physical Action Journal to recognize patterns, not to develop an absolute or rigid theory of your mind.

Get Specific. What's Underneath?

One issue that often comes up is the impulse to use verbs that aren't specific enough. I tend to ban the word "convinces" from lists of physical actions. It isn't that we never try to convince people. It is more that, in some respects, every moment is some form of convincing. Leaning on that action helps you avoid peering more deeply into the reason you did what you did. I also worry that, for most people, it has a particularly verbal connotation. When actors select it the rest of their body disengages.

Often these first attempts to keep a journal also result in a lot of repetition. There are probably physical actions that you play frequently. That said, if there are ones you use over and over every day, look for more specific and differentiated ways to describe what you're doing. Maybe you wrote "charms" multiple times a day. That might be a common physical action for you. It might also be true that some of those moments can more specifically be described as "wins" or "hooks" or "dazzles." It is good to start with the word that comes to mind, but pull on that thread and see if it unravels.

The last common issue is verbs such as "flirts." To make it work you need to add something. You have to say "I flirt *with* them." At first glance this isn't such a big deal, but it does have some important consequences. Remember from Chapter 3 that the addition of "with" allows you avoid truly testing the result in your partner. That little word makes the physical action about you, not about them. It is less an action and more a description of a set of behaviors. It encourages you to *act like* you're flirting. The more interesting question is what is under the behavior of flirting. What are you trying to see happen in them? There are many possibilities. Maybe the action is as simple as "seduces," but it could also be "taunts" or "tests" or even "shrinks" or "betrays." You can see from these physical actions that the *behavior* of flirting is really just one possible manifestation. It is the "what" or the empty "how" with no "why" to fill it. That's one reason you should always phrase the physical actions "I verb them" or "Name verbs Name." Doing that makes it much harder to select a verb that doesn't fit. You wouldn't write "I flirt them" or "Ky flirts Vi" because it doesn't make grammatical sense.

Take a moment now to rephrase the physical actions from your journal that are too literal, too general, or don't really fit the format. Now that you have some distance from these moments it may be possible to think back and quickly come up with more specific physical actions you played. Others will remain elusive.

Discover More Through Physical Playback

Poking around your initial observations generates a vocabulary to describe your behaviors and the needs that motivate them. As you wrote and rewrote your lists while recalling moments from your day, did you speak, whisper, or mouth the words you said during the original exchange? Perhaps you even fully or partially replayed the gestures or other behavior to help you understand what you were trying to accomplish. If so, you already intuitively understand that all actions are physical actions. There's evidence that taking on the same physical shape you had during an event can help you remember that event.[7] It is easy to focus on just the words we use to express our needs, but doing so ignores how the events actually happen as well as all the information listeners and viewers get from the rest of our bodies. We know that gesture is a valuable component of how we form thoughts and how listeners understand what we mean.[8]

There's so much to gain from better understanding how we communicate meaning physically and how the information our bodies send to listeners supports or undermines our words. Using a vocabulary of action verbs gets you thinking about human interaction as a physical event with real visible manifestations. This is a vital shift for you as an actor. If you begin your process acknowledging that thoughts are physical you can harness that as you explore a role. Looking at your journal, find some moments when you wrote what happened but didn't manage to select a verb for the physical action. Find a room or studio where you can experiment a bit further.

Physical Playback

1. *Select about six moments you couldn't decide on a verb* for (or, if there are none, select actions you simply want to explore).
2. *Look over the notes you took about what you said and did.*
3. *Now, think back to that moment and try to recall what you can of the event.* What was the room/location like? What were you or the other person wearing? Anything else about the environment or their appearance you can recall? Close your eyes for this step if you like.
4. *Put your body in a similar shape to that moment.* Were you standing or sitting? How were you turned? Can you remember the way limbs were crossed or if you were holding any objects? Were you walking?
5. *Where was the other person?* Were you looking at them? If not, what were your eyes focused on? Open your eyes, if they're closed, and replicate this.
6. *Now play back that moment.* What was the verbal and non-verbal embodiment of the event? Try to keep it as close to your memory as you can.
7. *Do it several times.* If you seem to recall new elements, incorporate them with the repetitions. It could just take a few seconds or maybe it was a longer moment.
8. *If it is hard to catch it all at full speed, slow it down.* This may seem silly if the moment was relatively unimportant, but sometimes that makes it harder to identify the physical actions.

9 *After a number of repetitions playing the moment back, see what verb fits best.* Do the words you spoke offer any clues? Do the position of your body in the room, your gestures, or other non-verbal elements help identify the impulse underneath?

10 *Repeat steps 2 to 9 with the other moments you selected.* Remember, it doesn't have to be right. It is your best assessment.

After you've done this with several moments, pause to reflect on the experience. What elements of the way it was physicalized come back to you by actually doing it again? What are the holes in your recollection? Can you tell what you're inventing to fill in those gaps? How often are your gestures and words in agreement or disagreement? Do you have physical habits that undermine or contradict your words in certain settings? If you played the same physical action multiple times, did it come out differently with different people? Can you identify parts of your body that you include or exclude in each version? It is fine that your memory isn't perfect. In reality, even if you think you remember it perfectly, you don't. Notice what you can, the gaps in your observation, and use that information to observe more specifically moving forward. Do you find pleasure in replaying some of these actions? Did any reveal new information about the relationship or the way the other person experiences you? Are there some moments where you wish you played a different action?

Let's reverse that experiment now. Looking back at your journal, pick several moments where the word for the action came to mind more easily and try the Physical Playback experiment with those moments. You can jump right in or begin by writing what you can remember about how you played the action in the column next to it. Pick a wide selection from different days and in different settings. If you used language to express yourself, try to remember if there were also non-verbal elements to the action. (There were, however small or unmemorable.) Often people only remember the most obvious element of how they expressed the physical action.

After doing that, keep asking questions about the results. Do you use the same verbal expression to accomplish many different actions? Are there certain gestures or other non-verbal behaviors that you use in many different ways? Are certain types of actions more verbal or non-verbal in your expression? Does the non-verbal expression embody an idea you have about the world? Maybe when you replayed some moments you were more forceful than in real life – releasing more strength of impulse than you felt you could when the moment happened. We often replay meaningful moments in our minds or rehearse challenging conversations in advance. Give that impulse greater physical liberty and discover what new information it provides. Taking on the body position helps you remember the event.[9]

Three Types of Actions

Putting language to a complex experience always feels like estimation. Despite that, there are very clear relationships between the way we think about words and how we perceive the world. The verbs themselves offer insight into how you experienced the event or the impulse underneath. Some actions have an obvious physical version. If you played "slices," maybe you used language to do it, but you also have many non-verbal expressions of that action. The most obvious is cutting something with a sharp object, but you might also imagine a crowded sidewalk with people walking shoulder to shoulder and you passing cleanly between two of them to zip past. That's a way to play "slice" with your entire body that mirrors the way you already think of it in a literal way with a knife. We'll call these actions with an obvious physical version "primary actions."[10] They already have an embodied way they exist in space, but you don't only use them in that way. If you have a strong desire to hug someone but can't, you might hug them with your words. The embodied desire is there, but the moment does not allow you to express it in the obvious physical way.

Some verbs don't have embodied versions that are as literally expressed, but they have lots of behaviors associated with them. These "conceptual actions" include words like "soothe," "seduces," "worships," or "rejects." Reading those words you can imagine verbal and non-verbal ways to express those impulses. There isn't a singular or iconic physical version of it, but there are several easily identifiable ways to play those actions from various situations in life. We now know that these ideas and their physical form are related. In fact, as Rick Kemp writes in *Embodied Acting*, "Cognitive science now shows that conceptual thought and physical activity frequently share the same neuronal pathways in the brain."[11]

The third main category is "metaphorical actions." For example, if you lust after someone you might play the physical action "devours." You aren't literally devouring them, but we have a link between desire and hunger in our brains that makes the ideas feel linked. This opens up a whole world of additional actions. Cognitive linguist George Lakoff and philosopher Mark Johnsen, authors of *Metaphors We Live By*, did tremendous work uncovering how we use language with embedded physical ideas present. Some of these metaphors are so ingrained that the discovery that they aren't literally true is amusing and surprising. For instance, "difficulties are burdens" or "intimacy is closeness" are two metaphors that fundamentally shape the way we think about the concepts. The ways we speak, think, and act based on the first word are deeply influenced by the second word. Another example with a great physical manifestation is "Having Control or Force Is Up; Being Subject to Control or Force Is Down."[12] They cite several examples of this in common phrases such as "I am on top of the situation … [They're] at the height of [their] power. [They're] in the high command. [They're] in the upper echelon. [Their] power rose … [They are] under my control. [They] fell from power ..."[13] In investigating the source of this deeply embedded metaphor they note that "Physical size typically correlates with physical strength, and the victor in a fight is typically on top."[14] These connections weave into our communication from an early age and are ingrained in our physical sense of the world.

Your Physical Action Journal may contain some of these metaphorical actions. Perhaps you played "elevate" at some point when you were trying to give someone more power or control. Maybe you turned on a light in your living room when your roommate was struggling with homework and thought the physical action was "inspires" – using the metaphor of intelligence as a light source. It isn't important right now that you thought of the metaphor when doing it. These ideas are so much a part of how we understand the world that we cannot help but employ them. You can play all three of these types of physical actions: primary, conceptual, and metaphorical in your work as an actor and they may already be in your Physical Action Journal. You may also realize that one of the types of actions is completely missing or your difficulty in naming physical actions for certain moments was because you didn't realize you could use verbs in one of these ways.

Take a look at the second version of the list of verbs at the back of the book. You'll notice they're grouped by type. I put the words "massages," "charms," and "thaws" in bold. These words each fall under a different category of action, but all might accomplish a similar goal in your life. If someone is upset with you and you're trying to get back in their good graces, you might have a conversation where you play one or more of those actions. They each aim for similar results in your partner, but likely produce a different way of playing the action within you. These words also evoke a different sense of the person who plays them. What kind of a person "thaws" their partner versus someone who "charms" them? Does it speak to the seriousness of the conflict? Does it generate ideas of physical behavior that might accompany the action? The words you choose unlock a whole series of options, even if they seem like synonyms. Use this list for the following experiment.

Switching Action Categories

Looking at your Physical Action Journal, ask a few questions about the categories of actions:

1. *Do certain types of actions (primary, conceptual, or metaphorical) show up more with certain groups or people?*
2. *Are there types that aren't in your journal at all?*
3. *Now that you understand these different categories, could you find better words to describe some of the physical actions you played by switching categories?* Maybe you wrote "loves" but

"warms" feels more specific and accurate in terms of what you did. It also uses the metaphor "affection is warmth."

4 *As an experiment, select several primary or conceptual actions and turn them into metaphorical actions.*

One category of action isn't better or worse than another. For now, you're practicing finding names for the ways you embody impulses. You might have strong habits about how you think of yourself in relation to others that make certain physical actions or types of actions easier to imagine doing. You may have judgements associated with certain primary actions, concepts, or metaphors that keep you from realizing that you play them. As you refine your Physical Action Journal find words that excite you by how specifically they name the experience of a moment. *A great physical action is one that seems to capture, in a single word, the impulse you had and the way you embodied it through verbal and non-verbal communication.*

You Can't Always Get What You Want

Some physical actions might be part of multiple attempts to achieve the same basic thing. In fact, what keeps an audience watching is often someone trying to get something over and over until they succeed, give up, get something else, or die trying. When you look back at your Physical Action Journal, take note of any times you struggled to get your point across, your need was rejected, or when someone misunderstood your intended action and you had to try again. Did you note several physical actions you played in that exchange? Or did you just end up writing the final one you tried?

It's useful to reflect on the actions you *first* attempt and then how you escalate or alter your tactics when you fail. Take a moment to go back and see if there are any physical actions you want to add the journal considering that it might take many attempts to land on a successful action. For your later work on character, there is substantial value in remembering the series of physical actions you play when a first attempt fails. If you tease someone a bit and play "jostles" but the action you get from them in response is "dismisses," then maybe they didn't take the joke well. How do you fix it? Some people double down and play "tickles," others retaliate by playing "wounds," and still others might try to fix it by playing "cheers." The shift from jostles to any one of these second physical actions reveals character and relationship.

You now have several pages of early observational notes about how physical actions operate in your daily life. Hopefully you already see that each interaction with another person is a process of playing and receiving actions based on your needs in the moment, your personal history, and your specific history with that person. Your notes also reveal that you don't simply play these actions with language. Everything from eye contact to bodily position, behavior, and more integrate into the expression. In fact, words often reveal less about our true actions than how we say them and what we do with the rest of our body.

Your Physical Action Journal should be full of cross-outs and rewrites and notes where you found a better verb through physical playback (changing the type of action between primary, conceptual, and metaphorical), got more specific by eliminating ones that were too general, or added ones when you made multiple attempts toward the same goal. Now that you collected these notes from the field, let's uncover how your physical actions operate as you move from relationship to relationship and role to role in your life.

Notes

1 Benedetti, Jean. *Stanislavski and the Actor*. New York: Routledge, 1998, p. 3.
2 Clark, Andy. *Surfing Uncertainty: Prediction, Action, and the Embodied Mind*. Oxford: Oxford University Press, 2016, p. 3. DOI: 10.1093/acprof:oso/9780190217013.001.0001.
3 Ibid., p. 295.
4 Gallagher, Shaun. *How the Body Shapes the Mind*. Oxford: Oxford University Press, 2005, p. 27. DOI: 10.1093/0199271941.001.0001.
5 Benedetti, Jean. *Stanislavski and the Actor*. New York: Routledge, 1998, p. 2.
6 Lutterbie, John. *Towards a General Theory of Acting*, Cognitive Studies in Literature and Performance. New York: Palgrave Macmillan, 2011, pp. 111–112. DOI 10.1057/9780230119468.
7 Dijkstra, Katinka, Michael P. Kaschak, and Rolf A. Zwaan. "Body Posture Facilitates Retrieval of Autobiographical Memories." *Cognition* 102 (2007): 139–149. DOI: 10.1016/j.cognition.2005.12.009.
8 Tversky, Barbara. *Mind in Motion: How Action Shapes Thought*. New York: Basic Books, 2019, p. 124–130 and 131–134.
9 Dijkstra, Katinka and Rolf A. Zwaan. "Empirical Support for Memory and Action." In *The Routledge Handbook of Embodied Cognition*, edited by Lawrence Shapiro. New York: Routledge, 2014, pp. 299–305.
10 Kemp, Rick. *Embodied Acting: What Neuroscience Tells Us about Performance*. New York: Routledge, 2012, p. 210.
11 Ibid., p. 99.
12 Lakoff, George and Mark Johnsen. *Metaphors We Live By*. Chicago, IL: The University of Chicago Press, 2003, p. 15.
13 Ibid.
14 Ibid.

5 Creating Palettes

Your Favorite Colors

Now that you started cultivating curiosity about your own physical actions and refining the observations from your journal, we can start to explore what set of actions you most often play by habit. I use the metaphor of a painter's palette, calling this your Primary Action Palette. The painter might have hundreds of colors sitting on the shelf available to use, but the ones already on the palette in a moment are the ones they can use most readily. In addition, there are likely to be a few the painter really enjoys using that gain a near-permanent home on the palette. Your Primary Action Palette is a bit like that. The more you use certain physical actions and the more they succeed in getting your needs met, the more likely you are to keep using them. They become a core part of your personality and how we think of you. Soon, they become habitual and you aren't even aware how often they come up. It is important to know your own action palette so that, as you work on a role, you understand which of your own habits you want to include or exclude. Without that understanding you're more likely to simply make every role's palette just like yours

Primary Action Palette

1. *Look back at your Physical Action Journal and circle the actions you used most often during the week.* If there are no repeats, are there some actions that seem to describe a category you use a lot (like "supports" or "halts" or "teases," for instance)?
2. *Make a new list of about 10–12 of these actions to create your Primary Action Palette.* If you are still struggling, look for ones that you think others would use to describe your regular behavior or maybe ask a trusted friend or partner to review the list.
3. *Be sure the selection represents you in various social environments of your life* – home, work/school, with strangers, or other groups that are a major part of your regular social interactions.

4 *Read it aloud.* Does that sound like you? Do you have an opinion or judgement about the person that palette describes? What person comes to mind when you say those actions aloud?
5 *If it seems entirely unlike you, look back on your list and switch out some actions that offer a more complete or accurate view of you.* Be careful not to exclude or eliminate actions just because you don't like them or feel judgemental about the way you might be viewed playing them.
6 *Do the Physical Playback experiment from* Chapter 4 *with a moment from the week for each of these physical actions.* Explore the way they manifest throughout your body. How does it feel to do them? Can you sense why them come up so often?
7 *Now, memorize that list.*
8 *During the next week, use the Primary Action Palette as a sounding board for your experiences in the world.* Review the list once each morning and once every evening.
9 *Keep your Physical Action Journal with you and make new notes about when you use them, with whom, and what you observe about the verbal and non-verbal expression.*

With your mind primed with these words, do you recognize the frequency with which you play these physical actions over the course of a week? What are all the different verbal and non-verbal ways you express them depending on the people and setting? Living your life with an action palette in mind generates discoveries about how others perceive you. It's okay to refine the list as the week goes on. Certain words turn out to be inaccurate. You might decide a different word is a better fit for what you are actually doing in those moments. You might also determine your palette is less varied than you thought. In reality, the majority of your interactions might be covered by about five physical actions.

Perhaps, situation to situation, the actions you play are so different that a single palette simply cannot accurately capture the character of you. That's an important discovery, and one we'll explore a bit later. Hold onto this real-world complication for your work as an actor. Too often actors develop a simplified version of their character's personhood. They distill them down to a few basic behaviors. Discovering your own shifting self, environment to environment, is an important reminder of how interesting and vital those shifts are for audiences and storytelling.

Expanding the Impulse

Let's dig into the impulses these words describe. So far you have asked why you did what you did by keeping a Physical Action Journal about those impulses toward others. You have investigated what and how you did it by exploring the verbal and non-verbal ways you played those actions. In this next experiment you'll take the physicality you observed and make it stronger to uncover the nature of the underlying want or need. This offers insight into the physical actions and reveals new ways to play them. It also opens up the possibility of renaming actions to better match the strength of your desire underneath.

Expand and Contract

1. *Select a physical action from your Primary Action Palette.*
2. *Playback the action as you recall doing it.* That includes any words you used, gestures, or other movement around the room, etc. Really try to include everything you can recall.
3. *Keep it brief.* Focus on the words and behaviors from the actual event that best capture the moment the action was most clearly and strongly played. Just a few seconds is plenty.
4. *Repeat that over and over so it becomes a short loop of the same behaviors and words (if you spoke).*
5. *Begin to expand the size and scope of the physical form.* For instance, if it is a tiny rotation of your wrist, could it turn further? Could the impulse start from somewhere else in your body? Perhaps your chest, or pelvis, or even your feet. Find the parts of yourself that are involved and make that more visible by strengthening and expanding their movement.
6. *Let this expansion open new ways to physicalize it.* Maybe as the original gesture expands it makes you want to take a step or jump or curl into a ball. Move beyond the parts of yourself that are already included.
7. *As you keep looping the physical action and expanding the non-verbal elements, let the verbal elements shift, too.* Do you say more, or do the words change? Perhaps you just end up making non-word sounds. Maybe a word in another language you speak feels like a stronger expression. Sometimes it helps to say the action verb itself to support your exploration.
8. *Keep expanding this loop until you engage your entire body (including your voice) in the strongest and biggest possible version of this physical action.* It is okay to keep changing and exploring with each loop. It shouldn't be "realistic" or anything you'd likely do in life. It is an attempt to embody the impulse as completely and unselfconsciously as possible. Let every part of your body help communicate this one action.
9. *It helps to keep increasing how important or high-stakes the action is.* Even if the moment in life wasn't like that, imagine an

incredibly important circumstance to help the strength of the need match the scope of the physical engagement.

10 *Don't let it become empty or hollow movement.* The impulse and desire to play that action must stay strong and alive.

11 *Once you reach the most extreme version you can find, immediately snap back to the original version but keep the need just as strong.* Try to rediscover that original verbal and non-verbal expression.

12 *Do that original version a few more times.* Can you feel the strength of that bigger impulse wanting to come through? How effective does this version feel now? Does it still get across the action you intend it to?

When you first began to expand the physical action, the early repetitions were probably still something you could reasonably do in life. Maybe you went from casually pointing at someone to really extending your arm as you pointed and squinting as you tighten your jaw. That would certainly be "realistic," even if you weren't that big (or confrontational) in the actual moment from your life. When did the size of the action cross over into unacceptable or unrealistic for you? When a certain part of your body became engaged? When your voice reached a certain volume? It is important not to assume that how you normally choose to play an action is the only way available. When you need something and try to get it in life you are calibrating to the other person, the setting, and ideas about yourself. Test those assumptions in these experiments in order to better understand the impulses you squeeze and shape before they come out. Try this exercise with several physical actions from your Primary Action Palette. You may discover that your impulses are stronger than you acknowledge in day-to-day life or that there is actually a different physical action under some of your impulses that you simply couldn't recognize without exploring a more expanded physical form. If that's true, feel free to change the verb you wrote down. It's okay to let the experiments change earlier ideas.

By expanding and contracting actions you discover that most people prematurely reject a range of expression that actually feels and looks perfectly realistic and believable. Sometimes we exclude the use of certain parts of our body for fear it communicates an action we don't intend. Other times our upbringing places rules on what is and isn't acceptable behavior so we suppress or cannot access forms of expression that other people use. The most expanded embodied version of an underlying action impulse might not be usable in performance but there are elements of it, or steps toward it, that are far more playable than you realize.

Having done this experiment, can you think of people in your life who allow more visible embodiment than you do? Others who use less? These factors (and how different they are from you) influence your point of view about them and the way you receive the physical actions they play. It's part of their character as you experience it. Don't assume that every character you play embodies actions the way that you do. Different mediums and types of material also demand differing kinds of embodiment, of course. Just think of dance. There are moments of profoundly experienced and expressed actions that utilize embodiment in "unrealistic" but achingly truthful ways. As you work on a role in Part 3, you'll do a series of experiments to develop greater variety for embodying actions in unfamiliar ways.

Who You Think You Are Not

Let's switch perspective. As many physical actions as you played in the past couple of weeks, there are thousands more you did not. The actions you exclude from your palette also tell a story about who you are and how you want to be seen. Take a look again at the short list of verbs at the back of this book and see if there are any that you *never* play or that the thought of playing elicits fear or dread. Maybe actions like "seduce" or "destroy" fit that description. For some people actions like "cuddles" or "adores" are too vulnerable or embarrassing. Sometimes we don't play physical actions because we tried them before and the results were negative. Are there physical actions like that for you?

Perhaps you also have judgement about the actions you see others play and don't want to be like them. They could simply be people you know in passing, but you may also have actions you associate with parents or other family that you very much want to be like or want to avoid comparisons to at all costs. On the other hand, some physical actions are so unfamiliar – so lacking a model from your life – that you cannot imagine how they would work no matter how much you want to play them. Often the actions people avoid are on the extreme ends of a spectrum. In life, it is easier to create a more and more narrow palette of tried and true physical actions that get a good response and you have lots of practice playing. In fact, trying ones outside that list might even damage relationships because of what people need or expect from you. Your work requires you to have a broad palette readily available for the circumstances of any project and role.

The Anti-Palette

1. *Make a list of 5–10 actions you never play or are afraid to play.* You can simply think of these, or you can use the short list at the back of this book.
2. *Why are they on your list?* Have you ever played them? What was the result? What do you think of people who play them?
3. *Take some time to imagine playing a few of them with important people in your life.* Close your eyes. Really take your time. What

scenario played out in your thoughts after you initiated with that action? What event or action from them was required to motivate that action from you? What were the verbal and non-verbal elements you imagined for the action?

4 *Use the Expand and Contract experiment from earlier in this chapter to embody a few.* Begin with the imagined scenario and expand from there. Don't forget to allow verbal expression.
5 *After you reach the far edge of how expanded each action can be, contract it back down over several loops.* It is okay if the way you play it shifts. Use your discoveries.
6 *Don't just snap back.* What parts of your body feel less engaged as you contract the action? Does the strength of the impulse remain or does that contract, too?
7 *When you return to the "realistic" version, be sure to repeat that a few times to observe as much as possible about where in your body you experience the action.*

After you try that with several actions from your Anti-Palette, take some time to reflect on the experiment. What did you have to imagine in order to justify those physical actions? Did the scenario change as it expanded and contracted? Did other avoided or forbidden actions pop up as it played out in your imagination? Did playing these actions generate an emotional response? Did you notice anywhere in your body that there was an unfamiliar or unexpected muscular response? Was it fun to have permission to play these actions?

Imagining someone and seeing that person's image activates the brain in similar ways.[1] That may help explain why it is sometimes difficult to even pretend to do unfamiliar or undesirable physical actions with people we know. Maybe we fear it will give permission to actually do it in life. Maybe we worry that being able to imagine it is a sign of something wrong with us. Some actors express fear that imagining something terrible (a parent dying, for example) has some ability to make that thing happen. I suspect that some of these ways we resist imagining the unimaginable in our work is a result of how connected imagination and perception are in our brains. An *imagined reality* is, in some respects, reality.

You cannot be like everyone else in this regard. Actors must be fearless imaginers. In this process of better understanding who you are through physical actions, it is vital to understand who you are not or don't want to be so you recognize that resistance as you approach a role. Practicing and exploring unfamiliar actions make them more available to you in your work, expanding what seems possible to do with your body.[2] This doesn't mean you need to become a different person in your life, but it does mean that you benefit from experimenting with a vast number of physical actions you rarely play so that when the opportunity arises in a rehearsal or performance, you have a path that makes those impulses readily available. If there are colors an artist keeps in a box in the closet and never puts on the palette, there will be little impulse to use them to paint. After enough time, they might even forget they are an option. It is important to know what you're already doing so that you can expand your palette when your work demands it.

Expanding Your Palette

1. *Select a physical action from your Anti-Palette.* Choose something you wish you could play or something you believe is just not a part of your character.
2. *Find an opportunity play it in your life at some point during the next few days.* To be clear, I'm not suggesting you pick an action that would require you to physically harm people or destroy relationships for the sake of the experiment.
3. *Find something that seems doable, but outside your familiar palette and try it.* Did you change the action you first thought of when you learned you'd have to actually do it? Why?
4. *Who did you play the action with?* Why them? What was their response? Was the response an unfamiliar action from them?
5. *How did you play it (verbally and non-verbally)?*
6. *Did you really play the new physical action, or did you play an easier action?*
7. *What actions preceded it and followed it from both you and them?*
8. *Did you notice anything happen in your body in the moments before you played the action?* Was there an experience in your body that lingered after it was complete?
9. *Make some notes about the experience along with your Anti-Palette.*

This can be scary, but it can also be liberating. I encourage you to try it with other physical actions you never or rarely do. They don't have to be negative or combative actions. They might simply be outside your usual patterns. Check to see if your expectations about people's responses are correct. Try the same unfamiliar action with multiple people. Notice if your judgement about the action changes once you try it out. Does it provide you a new perspective on people who play this action all the time? Expand the available palette between you and others by bravely trying out some new physical actions and see what happens.

You Play Roles – We See Your Character

Looking at your Primary Action Palette and Anti-Palette, you already see that the actions you play and how you play them, moment-to-moment, are the result of a whole set of complex and constantly changing circumstances. It should come as no surprise to you that your interactions with others are full of history. Yet so often when actors come to work on a role they think about the life of a character, and especially their physical life, as a fixed set of traits. We talk about the posture of the character or the walk of the character or even the voice of the character. At its worst, this static view of a character's body pulls it apart from the history that created it, maintains split thinking about the mind and the rest of the body, and locks in behavioral choices that are unresponsive to the circumstances of scenes or the actions of partners in performance. We live in motion. Motion gives our minds feedback, which generates our next impulses. Taking action helps our minds to understand, assess, and navigate the world.

As you craft a set of physical choices for a role, it is vital to think of those choices as a part of interaction. You certainly have habits about walking or standing or speaking. They might even be quite consistent, but they aren't absolute. Maybe you cover your mouth when you laugh because you are embarrassed by your teeth, but not around family. You might know that you sit more on your left side because you keep keys in your right back pocket, but never on vacation. Perhaps you cross your legs a certain way quite consciously because it sends out a certain gender marker you want to express, unless you're in a space where everyone shares your gender identity. You can now see you have action habits, too. Perhaps you laugh a little right before you play an especially confrontational action to soften the blow. This collection of specific traits and habits are a significant part of the experience of you for others. When you do and don't engage those habits reveals information about your relationships. Making strong physical choices is important, but the best choices spring from a complex investigation of action and context.

When I speak to actors about their work, I generally refer to it as "playing roles." I think that keeps you focused on the doing. We play roles in life. Your physical actions moment-to-moment are in service of the role you're trying to play. On the flip side, what others (partners on stage or people in the audience) *experience* is your "character." They witness all those physical actions and, based on what they receive, put together an idea about your character. We speak about the character of others all the time. We talk about the character of politicians. Parents want to ensure their kids have good character. This distinction is important moving forward as we discuss the physical actions you play versus how they are received by others. It also helps when working on roles where you have a strong negative opinion about the person. There is great empathic value to focusing on the role the person is trying to play through their actions and let their character (and any judgements about it) be experienced by others.

Zoom Out

You've kept your Physical Action Journal, experimented with how you play those actions, and recognized some actions you exclude or avoid. Now step back and see yourself the way others might. Assuming the people you interacted with received the actions you played (or something close to what you intended), take a look at the "character of you" experienced by others. Does your Primary Action Palette reflect who you think you are? Does it reflect the way others seem to treat you? Does a different set of physical actions feel more like the real you? You're not completely the same in every circumstance, so the question is more general for the moment. Does this seem like a version of yourself that you recognize? If you find yourself having the response I described earlier – that you appear to be some manipulator always trying to trick people in your life – perhaps you'll take comfort in the origin of the word "person." It comes from the Latin word *persona*, which means "character in a play" or "actor's mask."[3] We've known for a long time that who we are is partly a performance for others.

Notes

1 O'Craven, K.M. and N. Kanwisher. "Mental Imagery of Faces and Places Activates Corresponding Stimulus-Specific Brain Regions." *Journal of Cognitive Neuroscience* 12, no. 6 (2000): 1013–1023.
2 Lutterbie, John. *Towards a General Theory of Acting*, Cognitive Studies in Literature and Performance. New York: Palgrave Macmillan, 2011, pp. 115–117. DOI: 10.1057/9780230119468.
3 Merriam-Webster.com, s.v. "person." Accessed August 1, 2018. https://www.merriam-webster.com.

6 Diagraming Identities

Who You Are Depends on Where You Are

Keeping a Physical Action Journal probably revealed notable differences in the actions you play depending on who you're with. While you have a Primary Action Palette that you use most of the time, there are certain circumstances or roles you play – even on a daily basis – that require other sets of actions. Sometimes this means employing an entirely different action palette, adding to your primary palette, or eliminating actions that are not acceptable or helpful with those people. As we discussed, no rigid group of physical actions reflects an entire complex person and it certainly doesn't make for very interesting characters. Then how do you build greater complexity in your work? You craft the ways you shift actions from person to person, group to group, and location to location. Watching those changes reveals story and character to your audience.

Sociologist Erving Goffman explored similar questions in his book *The Presentation of Self in Everyday Life*. He breaks the performance of self into two main parts: "setting" and "personal front." Setting is essentially the location(s) where the performance is given. Personal front is all the stuff we carry with us. It consists of two parts: "appearance" (which includes things like race, height, clothing, hairstyle, etc.) and "manner" (which includes everything that changes moment-to-moment, such as gesture and expression).[1] Manner is the visible manifestation of the physical actions you play. It is the "what" and "how."

A simple example might be someone selling electronics in a retail store. The setting is the showroom floor where they give the performance of "salesperson." The personal front includes their appearance. Is there a uniform? Are there requirements about grooming? It also includes visible traits that are or are not changeable, such as height, race, and gender expression. The second half of personal front is manner. *What* do they do and *how* do they play the role of salesperson? Do they seem professional in their interactions? Uninterested? Overly enthusiastic? Pushy? We generally try to make our personal front match the setting in order to fit in and meet expectations.[2]

Electronics salesperson happens to be a job I held for a while in high school. I wore a suit and tie and worked in a big retail department store at the local mall. Was I a completely different person selling those televisions and cell phones than at school or home or with friends? No. Much of my Primary Action Palette from the rest of life followed me to work. Customers experienced elements of my character through the physical actions I played with them, but there were other actions I excluded from that performance because they didn't match the setting.

I also added new actions because they fitted the role of salesperson better. Sometimes the days would be slow and there wouldn't be any customers for a while so I could drop those actions. Other times I had a break and would sit in the storeroom with my manager and chat. In these moments, the action palette I used was closer to the palette from outside work. Maybe not at first, but over time those co-workers became more like friends and saw me play physical actions usually reserved for that role. When a customer approached, however, it was right back to the action palette demanded by my idea of a salesperson.

I offer this example because, obvious as it may seem, it shows an element of performance in everyday life we engage in so effortlessly that it becomes invisible. *We all play different roles based on the setting.* We'll work in this chapter to experiment with how these different roles overlap, conflict, and ultimately blend to create a more complex and comprehensive understanding of the character of you.

What Are Your Group Identities?

Looking back at your Physical Action Journal, how easy is it to identify the major roles you play in the performance of yourself? You may have a few key words you use to give people a general sense of your identity. When asked to describe themselves many people list their occupation, gender identity, sexual orientation, age, race(s), ethnicities, or a variety of other factors. These are all group identities we use as a shorthand, even though we might not all agree on their meanings. Often people don't even realize they have a certain group identity unless it is a particular point of pride, something that others use to refer to them, or a way they experience feeling different.

Group Identities

How would you describe yourself thinking about each of these possible elements of identity? Write an answer for each.

- Age
- Education
- Socioeconomic class

- Religious beliefs
- Family roles (child, sibling, step-parent, lover, spouse, orphan, etc.)
- Sex (biological body parts at birth)
- Gender identity (maybe same as the body parts you have, but maybe different)
- Gender expression (how it shows to others outwardly in general or day-to-day)
- Sexual orientation (who you're sexually attracted to, or not)
- Citizenship
- Races and/or ethnicities
- Skin color (it may not match how people expect your race/ethnicity to look)
- Disabilities
- Physical size and shape.

This list is a good start because everyone has all of these group identities, whether they think about them or not. Notice which ones were easy to name and when you struggled or defined yourself as not-something. There are likely other very strong group identities you hold that you choose for yourself. Those could be a job you have, or being a veteran, or a club you belong to, or a sports team you support. It certainly takes a strong group identity to spend hundreds of hours a year watching a sports team and experience tremendous emotional responses if they win or lose. If you're reading this book you probably also have the group identity of "actor." What are some other important ways you identify that aren't offered as categories here? List a few along with the ones from above.

No matter how many of these you think about regularly, there are likely group identities that you never consider. Maybe your identity is considered "normal" by the society or country you live in and you are able to go through life not thinking about it. Some of these are visible and others are invisible. With any of these group identities or roles you play, other people who share that identity become "in group" and others become "out group." This sense of belonging (or not) has substantial impact on how we view others. Studies suggest that even the parts of the brain you use to *think* about other people changes depending on how much you identify with them.[3]

Code-Switching

With all of these group identities as part of each of us, we inevitably find moments where one takes precedence over another, seems more advantageous than another, or where one seems like it would be received negatively by others. The term "code-switching" comes from linguistics and describes the event of someone "switching from the linguistic system of one language or dialect to that of another."[4]

The term is also commonly used to refer to people altering their speech, language, or behavior depending on whether or not they are in a space where other people share their race or ethnicity. For instance, if you're African American or Black, you might use a voice you perceive to be more "White" in certain settings than you do with family or friends who also identify as Black. Maybe you suppress or add other physical or gestural traits, too. This self-preservation can be so practiced and habitual that people barely know they're doing it. Other people are acutely aware of the shifts they make to meet expectations or to avoid possible negative consequences for seeming different. Still other people feel that their regular behavior doesn't match expectations for their identity and they feel on the outside of *every* group. Many of you who identify as part of a minority group don't need an explanation of this at all given your lived experience. For others of you, this idea might be new. Elements of this code-switching happen to all of us, even if the risks for failing to do it are far smaller, or the perceived need to do it less immediate, than in the example above.

Think about the various roles you play in different settings in your life. How do the group identities you listed above impact the personal front (appearance and manner) you use? Are there certain settings and roles that require you to code-switch because of elements of your appearance or other factors that you cannot change? The physical actions you play might be significantly impacted by group identity and the need to code-switch. Consider my retail sales job from earlier. If you're a white woman selling electronics in a store with predominantly white customers, you might play certain physical actions more than I did to overcome potential bias about women and technology. Perhaps there are actions you eliminate that I could continue to do without fearing negative consequences. If you're a black woman doing that job in the same store you might make different or additional shifts given the multiple group identities at play and the personal and societal experiences that surround your relationship to that space. The job title is the same, but the action palette available or required might be very different.

Recognize Your Shifting Palettes

If a single action palette cannot fully express your character, then we must find a tool to express the adjustments you make and when you make them. First, we need to name those key identities in your life and get more specific about how you shift physical actions depending on settings, roles, and group identities. Take a look back at your Physical Action Journal to do the next experiment. It is worth continuing to add observations to the journal each day, even as you read the rest of this book. Your skills in observing yourself and naming your physical actions improve both with practice and as you understand how actions work.

Grouping the Actions

Take your Physical Action Journal and rewrite it into several smaller lists described below. Leave space to write next to each action for the next task. Some or all of the categories below will work for you, but feel free to add or subtract based on the way your own life is organized and, perhaps, compartmentalized. It is ok to reuse actions for multiple lists.

- What are the physical actions you played with family (however you define it)?
- What are the physical actions you played with friends?
- What are the physical actions you played at work or school?
- Are there physical actions you played with people who share a group identity with you (i.e. race, religion, gender identity, nationality, etc.)?
- What is one other group identity that could have its own list?

Take a moment to review this reorganized version of your Physical Action Journal. How many actions appear on multiple lists? Are there roles you left out? Could some of the lists be combined because they're actually one role you play? Are there people that see you in multiple roles in your life? Essentially these lists are unique action palettes you use in each of these settings.

Visualize How Your Palettes Interact

With all that going on every day it's a wonder you can keep track of yourself. All of these different action palettes combine to create your character. Sometimes you simply play one role, walk away, and play the next. Think of leaving work and arriving home to your family or roommates. Other moments are more complex. There are circumstances where you need to play two roles at once. Perhaps you're at work and a former romantic partner comes in that you haven't spoken to for a long time. Usually these worlds are separate. Here they are colliding. This attempt to keep apart different versions of ourselves is referred to as "audience segregation."[5] That simple example of an ex coming into work sparks the imagination toward theatrical scenarios. Scripts are full of moments when someone must negotiate between two roles, two palettes, two selves. It's the reason so much drama exists when we discover someone leads a double life. We want to think that the action palette we experience is the same one people use all the time – that it is the *real* them. It is, and so is whatever else they're doing that we never see.

In your work, thinking in terms of physical actions you include and exclude in different settings is a valuable way to keep a sense of your whole character while allowing the audience and collaborators to experience the shifts in actions that also happen in life. Using the lists you created by grouping the actions above, let's create an Action Palette Diagram. It is kind of like a Venn diagram. You might remember those from math class. They are a series of (sometimes) overlapping circles. If you get confused as you work, refer to the example right after the instructions as a visual aid.

Action Palette Diagram

As you work on this, use pencil so you can redraw and rewrite as needed.

1. *On a blank page, draw a circle (taking up about ¼ of the page) and label it "family."*
2. *Inside that circle, write 4–6 verbs from your Physical Action Journal that you just play with family.* Feel free to define family however you see fit. It may be a single person. Maybe they aren't related to you but serve that function in your life.
3. *Now add an overlapping circle and label it "friends" or "work/school." Write 4–6 actions you play with just this group in that circle.*
4. *Repeat this, adding 1 or 2 more overlapping circles. Label them for the other major group identities in your life.* This could be a specific friend group, or maybe it is related to a group identity based on race, gender, or shared first language(s).
5. *In each circle, try to write the physical actions you play most often with that group.* Use your lists from the Grouping the Actions experiment above to help identify them.
6. *Now look at your Primary Action Palette and make sure all of those actions are present on this diagram.* Do they all go in the center where all the circles overlap? Do some go in the space where just two circles overlap? Maybe some occur only in one circle, but that represents a major part of your day-to-day life.
7. *Step back and take a look at this diagram. Make sure the physical actions are in overlapping spots when appropriate.* Erase and rewrite and move as many words or circles as you need to. It's fine if it is messy. Below is an example of three palettes and how they might overlap.

You can see from the example on the following page how they play "adores" and "needles" regardless of the setting, while they reserve other actions for specific situations. They play "confronts" with friends and their partner, but not their family. The totality of their *character* as a person includes all three of these intersecting action palettes, even though all actions aren't available at all times.

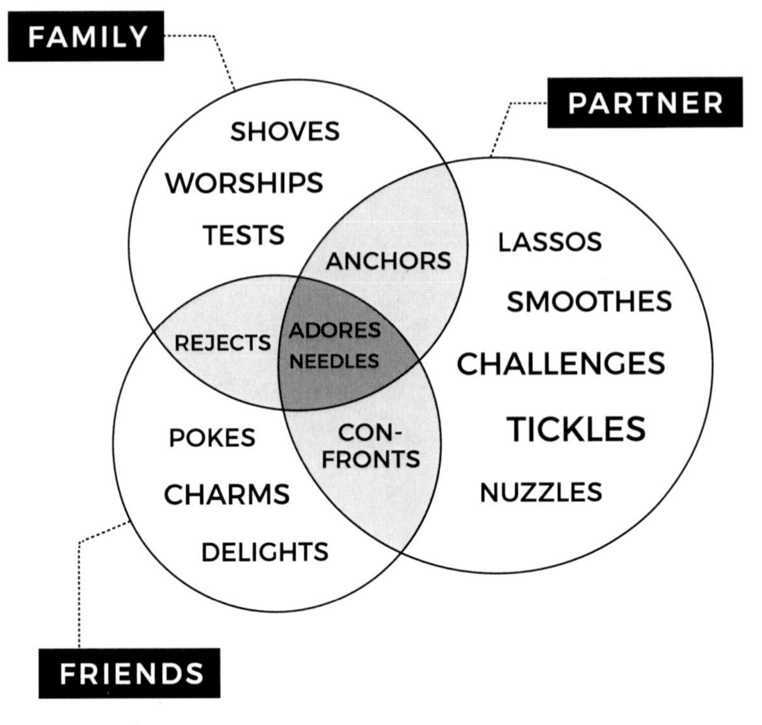

Your Action Palette Diagram

Looking at your own physical actions organized this way, think about what is invisible to certain people/groups about your character. Can you understand why you don't include those actions with them? You can keep adding circles for many different settings in life, creating a much more complex diagram. Standing back from this diagram of your action palettes and viewing yourself in this way may also reveal why people respond differently to you in different settings. Does one of these palettes seem more like an essential version of yourself? Your most comfortable version of yourself? The one you most *want* to be?

The Action Palette Diagram might also offer unexpected glimpses about how you choose to meet expectations about your roles as a member of a family or group without realizing those pressures exist. The roles of parent, partner, best friend, barista, social justice organizer, or any other named roles, hold expectations that you learned from others and the world around you. Your physical actions are situational. There are expectations that we all try to meet to foster the impression that the role we are currently performing is our only role or at least our most essential one.[6] As you can see from your diagram, we do this partly by keeping some of our actions outside the palette for that role.

Perhaps you have tried to buck this system and shake off expectations about how someone in a role should act. You likely found it a hard road. Or maybe that visible resistance is an important part of your identity. If one of the action palettes you included in the diagram was for a marginalized group identity, you likely also recognize how that group identity impacts the other palettes and the actions you play in those settings. How do the physical actions you chose for the other palettes serve as a response to expectations about group identity? This is related to our previous discussion about code-switching.

As you can see from these interlocking circles, no part of your life is completely distinct. There are clearly defined roles, but also areas of overlap and ways that one part of your identity impacts and alters the others. Now that you're starting to understand what actions you play when, the next question becomes "*How* do you shape those actions to get what you need in different settings?"

Notes

1 Goffman, Erving. *The Presentation of Self in Everyday Life*. New York: Anchor Books, 1959, pp. 22–24.
2 Ibid., p. 24.
3 Banaji, Mahzarin R. and Anthony G. Greenwald. *Blindspot: Hidden Biases of Good People*. New York: Bantam Books, 2016, pp. 135–140.
4 Merriam-Webster.com, s.v. "code-switching." Accessed August 16, 2018. https://www.merriam-webster.com.
5 Goffman, Erving. *The Presentation of Self in Everyday Life*. New York: Anchor Books, 1959, p. 49.
6 Ibid., p. 48.

7 The Pasta Maker

Socialized Actions

Sometimes, while creating these diagrams for their lives or for a role, actors think they have completely separate palettes in each setting. That can happen. Perhaps the plot of a script is about someone leading a secret life or performing a false version of themselves. Or maybe in your own life you have very distinct groups of friends that don't interact. More likely, however, key physical actions remain as a part of your palette no matter who you are with. Rather than finding a different action, the question then becomes, how does *the way* you play those physical actions change? Looking back at your diagram, are you aware of how differently you play overlapping actions depending on the setting? Even actions that are part of that center section where all the circles overlap might have very different embodied versions from group to group.

Exploring how you embody the same actions depending on circumstances is exciting work for an actor. Perhaps it is a shift in the words you use. Maybe certain relationships allow physical contact to play actions. Sometimes a scene requires a particular activity (washing dishes, for instance) that provides an opportunity to continue to play actions on your partner through the embodiment of that activity. Anyone who has washed someone else's dirty dishes while the person was in the next room has played an action through activity.

I often use the metaphor of a pasta maker to describe the socialization process an impulse goes through from thought to embodiment. The strong impulse to action is the raw unformed pasta dough that goes into the top of the machine. The machine that squeezes and shapes the dough is all the rules of the world together with the setting, relationship, history, group identity and other factors. The pasta that comes out on the other side is the socialized version of that impulse that the world, your partner, and your audience see.

So many factors, both in life and for a character, impact the verbal and nonverbal ways the action comes out the other side. Too often actors focus on either the raw emotional content of the experience or the affect – the visible external signs. The risk is that neither of these keeps the focus on your partner or fully acknowledges the circumstances that shape the impulse. You must understand the machine, not just make the dough, or serve up the pasta.

So, what factors in your life shape the way you play actions? Take your Action Palette Diagram and draw a four-sided box around the entire diagram. On each side write some factor that impacts the way you play all of your physical actions. *How* it impacts those actions might change depending on the setting, but the factor is always present in some way. For instance, maybe you think your age, a disability, your immigration status, and your religion are all factors that shape actions no matter the setting or role. Check out the example below.

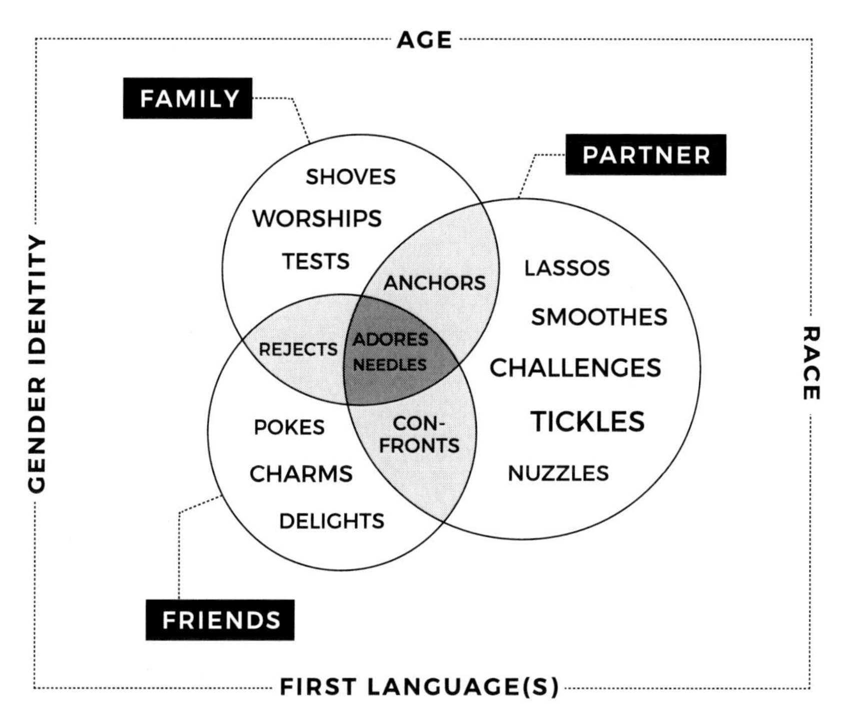

What Shapes Your Actions?

When you play physical actions, your partners see some of the taut relationship between the strength of your impulse and the demand to socialize it. Those strong action impulses bubble up and then get shaped by the box as they exit and become visible in the world. The result is still a wide range of expression. Sometimes the embodiment of our actions is in direct physical contact or relation with others. We hold open doors or hug or wag our fingers or play footsie. Other times our partner is only a witness to our physical behavior, but the action is still being played on them. For this next exercise, you'll take some of these raw, unformed actions and explore how you use everyday activities and behaviors to express a socially acceptable version of a potentially much stronger impulse.

Outside Behavior, Inside Action

1. *Select an action from your diagram in a spot where two or three circles overlap.*
2. *Recall a conversation or interaction with someone who is part of one of those circles where you played that physical action.* If you can't recall a specific instance, you can invent a reasonably possible one.
3. *Think of a simple domestic activity that takes some time and begin to do it.* Perhaps it is chopping vegetables, or cleaning the floors, or painting a wall. You can perform the activity in an imagined way. For instance, you don't have to actually paint a wall, but perhaps having a real paint brush would help.
4. *Experiment with playing the action through the activity.* How could you "adore" a person, for instance, in the room through this activity? Are you adoring them or are you adoring the object you're using? How would those be different?
5. *Could you do it without looking at them?*
6. *How strong can you make the impulse even though you might not be speaking?* Think of it like a bank shot in pool. The activity becomes a way to bounce the action to your partner even though you're not speaking to them directly.
7. *Now, try adding speech.* What would you say if they were in the room? Are there specific phrases you remember from the conversation?
8. *Does speaking alter your ability to use the activity to play the action?* Make it stronger? Weaker?
9. *Once it feels as strong as possible through this activity, begin to expand the outward physical expression of the action.* This is just like you did in Expand and Contract in Chapter 5.
10. *Let go of the realistic version of the activity and let more and more of your body become visibly involved.* Over some time, let it grow into something that is unrecognizable as the activity, but very much the fully embodied version of the action. If there are objects, it is okay to let them go – especially if they would be unsafe to use in this way.
11. *You can now direct it right at the imagined partner.* In this expanded form, feel the strength that comes from sending it to a target without the activity in between.
12. *Did you lose the speaking?* Could it help the strength of the rest of the embodiment?
13. *With the action expanded in this way, are those sides of the box you drew still present? How?* What does the fully expanded action look like when you feel free to express it in a way feels most released for you?

14 *Once you've gotten it as expanded as possible for a minute or two, shrink back down to the activity over 30 seconds.* Can you maintain the strength of the impulse even as you channel it into a more everyday behavior? Where in your body do you lose the sense of it? Where does it most easily stay clear to you? When do you shift your attention away from them directly and back to the activity as a tool of expression?

15 *Eventually you are doing the imagined activity again, but with the memory and experience of how strong the impulse driving that activity can be.* Try to visualize all of that strength and size being funneled into this much smaller expression – like a focused jet of water versus a wide spray.

Variation 1: You can also try writing a bunch of activities on scraps of paper, making a pile, and then write verbs on other scraps in another pile and pair them randomly.
Variation 2: Try playing the physical actions holding and/or using an object of personal emotional significance. Try one of no significance.

Notice what muscular engagement, thoughts, or internal sensations remained after you expanded and then contracted the action. Were you able to maintain some of the intensity? Did it remind you of certain circumstances from life? Sometimes, in the fully expanded form, elements of that box around your diagram become stronger or fade away. You may realize that molding the actions for the world makes you retreat from a part of your identity. When permitted to fully express the underlying impulse you might be freer to fully own that identity. In other cases, you may feel that the identity (and expectations around it) constricts your choices and the expansion process allows you to play actions without considering those factors. Perhaps both of those are true depending on the action. For instance, you might feel expectations to play certain actions in a way that matches expectations for your gender yet, when expanded, those actions find greater variety and liberty in their form. This might clarify some of the tension between the strength of the impulse and the socialized version people see.

It's good for this experiment if the activity and the verb seem quite mismatched. For instance, cutting vegetables and playing "destroys" seems fun, but easy. Cutting vegetables and playing "restores" is a more interesting challenge. Did you use primary, behavioral, or metaphorical actions? It's helpful to experiment with this exercise in each category. Cutting vegetables and playing "anchors" provides an exciting mix of experiences.

If you experimented with an object of personal emotional significance how did it alter the way you played the actions, or which actions you played? Did it impact the scenario you imagined? Did it limit your movement choices because you were worried about breaking it? Objects themselves can contain so much history and so much potential. If you play "dismisses" using a necklace the person gave you, it might be much stronger than the same action played through your phone. You endow some objects with a history that helps strengthen the physical action by their very presence.

My Feelings about Emotions

Often, we play physical actions through an activity or with words that don't match the verb because of how we want to be seen, how we fear people will respond, the rules and cultural values we were taught growing up, or because we don't even realize what we want from the other person. All of these factors are part of that box around your Action Palette Diagram. As you discovered, they shape and mold the way physical actions manifest verbally and non-verbally moment-to-moment. They determine the way the pasta comes out of the machine. Remember your Anti-Palette? This box might be the reason your Anti-Palette has the group of actions it does. You're always avoiding these physical actions, shaping your interactions so that the need for them does not arise.

When there is a large difference between the strength of your desire to play an action and the way it comes out, strong emotions are often the result. Think about the anger you feel when someone with lots of power says something unkind, leaving you to bite your tongue and smile because of their status. What emotion arises when you love someone deeply but know they are with someone else so you must keep your interactions free of obvious signs about your affections? This inability to fully release and play actions on a partner creates a tension that can produce an emotional response.

You'll notice that I don't mention emotions much in this book. When I do, it is often as an incidental result of other choices, physical actions, and events. A great deal of emerging neuroscience explores the nature of emotion and how it is embodied. Some methods in acting rely on this science to develop practical tools for consistent emotional response. There are certainly moments in performance where the circumstances or the text strongly suggest or even demand a specific emotional response from the actor. Many knees buckle at the sight of the stage direction "They sob." Many more wilt seeing their partner's next line is "Why are you crying?"

Some processes to produce emotion used by acting teachers and directors do, in fact, gain support from emerging neuroscience. Joseph LeDoux, at New York University, is a leading researcher in the field of emotions and the brain. His work, along with that of scientists such as Antonio Damasio and Paul Ekman, offers important discoveries about how the work of actors might capitalize on new knowledge about emotions. Recent studies help clarify how physical activity and emotion are linked, in part, through the experience of learning to use our body as it develops.[1] Even words related to emotion can activate emotional centers in our brains.[2] Many actors will wince at the thought of simply altering their body to create an emotion, but science reveals that the mechanism is similar. It is really our ideas and perceptions about how we experience emotions that make that path seem less "truthful."[3] Alba Emoting, for example, capitalizes on the ability to use controllable physiological states (body shape and position, breath, facial muscles, etc.) to generate an emotional response.[4]

While I think good performance can have a range of recognizable emotional responses, I don't think generating a large or cathartic emotional response is a useful area for significant focus for actors. In life, emotions are largely a response to circumstance and stimuli. We need something. We play actions to get that need met, constricted by the rules of the world and our sense of self. Our partner responds with actions that either meet that need or frustrate our attempt. That failure or success results in an emotional response. The size, shape, and nature of that response has everything to do with the environment we are in when it happens, the person we are interacting with, and how important that need is to us.

Emotions, especially strong and easily visible ones, are the *result* of something. I'm more interested in reliably creating that *something* since it's so much more a part of your performance moment-to-moment. If you play actions at every performance, they will tell the story, even if the feelings don't come along on a given day. The lack of emotional response in a character can also generate a profound response in audiences. Why? Because we put ourselves in the situation and have an empathic response to their struggle or their inability to express emotion.

Methods specifically organized to use newer knowledge about neuroscience to generate emotion are valuable in moments when the storytelling demands it or when an actor has a particular difficulty accessing an emotional response in imaginary circumstances. That said, I think a lack of emotional response in an actor (when one seems necessary for the narrative) is actually a useful diagnostic tool for you and your director. Why are the circumstances not producing that response? How could you alter the needs and actions and whatever else is directly about the events of the scene so that they more successfully produce that result?

Potency for Your Performance

You spent Part 2 taking a thorough look at yourself and your use of physical actions in life. The work of self-examination and observation isn't a single event. It is a continuous process of using curiosity to point attention at your interactions with the world. No doubt you can already see how directly this applies to playing roles and working with text. Your direct engagement with partners – attention pointed at the stimuli they provide – helps your mind stay in the moment and avoid slipping toward "lazier" predictive processes.[5] You must seek out what you have made habitual in your own life in order to practice the attention that will serve you in your work. Moshe Feldenkrais wrote about the need for this in his book *The Potent Self*. [6] Habituation costs us a capacity for spontaneity. Breaking this cycle feels deeply unpleasant. It doesn't feel organic. At first it feels far less spontaneous than simply continuing to respond from the safety of your habits.

I tell students that they are learning the beautiful discomfort of not-knowing. Parasitic habits rob you of your ability to take advantage of the moment you are in. You solve old problems. You might ask your self anxiously "If I eliminate these habits, who am I?" Observation of self doesn't turn you into a *neutral* person. Far from it. I don't believe in neutral. We are always moving, changing, shifting, balancing, and responding. Just standing "still" and putting your attention on your feet reveals a world of motion in supposed stillness. Instead of leading to neutrality, observation and curiosity lead to awareness. Awareness gives you choice. That ability to choose is potent. Potent behavior allows you to take action spontaneously without extraneous effort or corrective action in advance. It isn't neutral, it is a responsive and aware capacity to choose.

Part of this process of seeing yourself is seeing how you fit into the larger systems around you. Understanding the way setting and group identities impact you and others, and how we all interact is a vital area of exploration for you as an actor. All the roles you'll ever play have group identities. They have moments on stage where they are in-group or out-group. They likely have moments where they are forced to code-switch and that may profoundly impact their action palettes. Knowing this requires you to examine more about where your character comes from and how they see themselves to understand the rules of *their* world. Any culture has values integrated within it. Without that specificity, you risk making generalizations or stereotypes and simply drawing the character to your own limited experience. If you begin working on every role without examining your assumptions of normal, you miss a tremendous chance to understand why the character does what they do. You leave yourself no option but to assume the character has your same values, privileges, and marginalization. Perhaps there is a great deal of overlap, but likely there are some significant differences. This investigation should thrill you. You've spent the last few chapters looking at yourself and decoding your habits. You are ready to begin applying new embodied tools to a role.

Notes

1 Lutterbie, John. *Towards a General Theory of Acting*, Cognitive Studies in Literature and Performance. New York: Palgrave Macmillan, 2011, p. 143. DOI: 10.1057/9780230119468.

2 Moseley, G. Lorimer, Alberto Gallace, and Charles Spence. "Bodily Illusions in Health and Disease: Physiological and Clinical Perspectives and the Concept of a Cortical 'Body Matrix'." *Neuroscience & Biobehavioral Reviews* 36 (2012): 34–46.

3 Kemp, Rick. *Embodied Acting: What Neuroscience Tells Us about Performance*. New York: Routledge, 2012, p. 173.

4 Ibid., pp. 184–188.
5 Clark, Andy. *Surfing Uncertainty: Prediction, Action, and the Embodied Mind.* Oxford: Oxford University Press, 2015, pp. 260–261. DOI: 10.1093/acprof:oso/9780190217013.001.0001.
6 Feldenkrais, Moshe. *The Potent Self.* Berkeley, CA: Frog Ltd, 1985, pp. 6–13.

Part III

Building an Embodied Role

8 Mapping Their World

It may be hard to believe we've come all this way and are only now applying the work to imaginary circumstances. You're better positioned to use these tools on a role you might play and fully embrace the point of view of this imagined person because you have observed your own life through the lens of physical actions – uncovering the palettes and identities that form your impulses. As you saw from your exploration of how action palates change depending on the partner and the setting, nobody is the same at all times. The totality of these overlapping action palettes and how they shape interactions with other people are what audiences experience as "character." Character is received by the viewer, rather than created by the actor. Your job is to be in a continuous action loop. In Part 3 you'll discover and craft those actions in the service of the story and the world you build with collaborators. You'll work through an entire process of exploring a role from first read of the script through initial rehearsals. All of the work in this section can be done independently by an individual actor. Part 4 will offer ways to apply this work with partners or in groups.

For many actors, especially early in their career or training, there are few opportunities to work on big roles with a long arc over the course of an entire project. As you read this section, I encourage you to do the experiments using a role you'd love to play, one that is complex, and rich, and challenging. I discourage you from picking one that you know too well or have too many preconceived notions about. I especially discourage you from picking one for which there is an iconic performance you admire. The risk is that you end up working to repeat someone else's version of the role rather than truly discovering your own. It is best to select a work you have the script for and can refer to often as a companion for this section. If you're using this text alongside a project you'll actually get to perform, all the better. Have a single notebook for this process since you'll make lists, reflect on exercises, and even draw some diagrams. I advocate for a paper (or a non-typed digital) journal for the role, where you can write by hand. This is partly because it makes it easy to have everything in one place. It also is because writing helps us think about things in ways that are different than typing and helps us better remember what we wrote.[1]

While this section imagines you're working with an existing script, much of what we explore is easily adaptable to processes where an ensemble generates the script collaboratively, too. That said, whether it is a four-hour play or two lines of dialogue for an audition, most of the time actors begin with written information about the role that an author created. Most authors are incredibly thoughtful about what they put in the mouths of characters. You should assume great rigor by the author. The things that confuse you or seem contradictory should launch deeper investigation. Dismissal of the author is often the first refuge for avoiding the hard work of your job.

Zoom Out

Stanislavsky knew the vitally important nature of the first reading of a script.[2] It's an opportunity you can never get back and the only time you're truly innocent of what comes next. Essentially, it's your last chance to be an audience member and experience the pleasure and surprise of not-knowing. Often actors begin the process of working on a role focused on their own journey. You have a duty to the text and the story. Your role has a particular job in helping to tell that story. Understanding that makes it vital to resist placing yourself in the position of your own role just a little longer. You have a brief window of time where you know so little about the play. Preserve that innocence and gather information and perspective that you will lose as you, necessarily, begin to see the world and events through the eyes of your role.

For your first experiment with the text, you'll read the script and keep a running list of everything you expect will happen next and the unexplained behavior of the people in the story. These might be huge events or just little intuitions. Essentially, it is a script-reading version of the very first experiment in this book – letting curiosity lead to attention. This is what your audience will do when they watch your performance. Put yourself in their shoes to understand the journey of the storytelling before ego or fear or anything else clouds your process. If this is a piece you already love it can be a challenge to do this, but do your best to return to a place of innocence about the text or, if possible, read it through while working to recall those first impressions.

Read with Innocence

1. *Keep a notebook next to you the first time you read the script.*
2. *Make two columns. One for the page number, line, and/or event. The other is for the expectation or question it generates.*
3. *As you read, make quick notes about these items.* Maybe an unseen character is spoken about several times and you note "I expect NAME to enter any second." Or maybe something about an exchange between strangers makes you write "I expect them to kiss by the end of the play." It can also be things that don't make sense at

the time like "Why did they bring up the story about that vacation?" or "Do they both think they are high status in this room?"

4 *Work softly*. You aren't a detective seeking out secrets. Let your natural curiosity gently turn your attention into questions or expectations.
5 *Read it all the way through the way it would be viewed*. Take a break at any intermissions. Maybe get up and stand in line for your bathroom. Perhaps a moment of pause between each scene. Replicate the viewer's experience.
6 *When you get to the end, take a look back at the story of the text your questions and attention produced*. How many of your thoughts played out? How many questions stayed unanswered? Was that on purpose? When did you go down an entirely incorrect thought path?

Some of these responses are individual to you, but it is likely that many of the same questions or thoughts will go through the minds of an audience as they watch your performance. "Why did that person come into the room?" "I bet they're about to deliver the news." "Why do people keep mentioning that one uncle?" "Why so many silences between those two characters? Maybe they're having an affair." Some of these are also likely to be the questions and curiosities the author intended an audience to experience. You'll never have the benefit of this innocence again. In many ways, it is your job to help ensure the audience experiences what you just did. Keep this document with you throughout this process. Serving the intentions of the playwright through your choices should be your north star as you move into the rest of the work. When you get lost about how to proceed it is a useful place to return.

What Journey Are We On?

After that first read as an audience member, your second read should focus on the major events of the script. A major event is something that is necessary for the next thing in the story to happen. It might be new information that changes a relationship or the arrival of a new character. It could be a piece of paper delivered or a gift given. It can be obviously big events like a car crash or a diagnosis, but it can also be seemingly smaller things like throwing away a paper or not answering a call. Whatever it is, the story can't proceed if it doesn't happen.

Read for Major Events

1 *Start back at the beginning of the script*. Keep your notebook nearby.
2 *As you read, write down any time something happens that changes the course of the story*. This can be the arrival or departure of a

character. Maybe it is a secret revealed. It could be an agreement being made or someone sharing that they are in love.

3 *Be careful not to make everything too important or to skip genuine events.* Assume that every scene has at least one event, otherwise why would it be there? Look for the moments that launch the next piece of the story or spur the characters to take action.
4 *Make note of not just what is said, but any visible physical elements of the events.* Does it include entrances and exits? An object being transferred? A meal being eaten? A kiss?
5 *When you're done, read the list aloud.* Does this tell a summarized version of the story? Does it make sense how one thing spurs the next?
6 *If I read just these notes would I understand the essential plot? Are there holes? If so, go back and find the major events you're missing to fill in those gaps.*

These major events become a brief summary of the script for you to reference as you work on the role and in rehearsal. How many of these events involve the role you're playing? How much happens without you present? It is important for you to see how you fit into the larger narrative. If your goal is to help tell the story through these events, then the work you do on your role should support that mission. Down the line, this way of thinking about your job clarifies the physical actions and helps you see the best way to give the audience the experience you had during your first read. Sometimes it can also help you avoid falling back on actions you play habitually because it becomes clear they aren't supported by the events of the play and what the story needs from you in the moment.

Find the Edges of the Map

Even though the work of acting is an attempt to answer, in some small way, what it means to be human, not every world you inhabit in performance is a replica of our world. Writers craft worlds that are familiar in some ways and magical or absurd or heightened in others. One of the great pleasures of acting is getting to live, for a time, in these altered realities. Still, it's easy to see these differences as obstacles to truthfulness. It may seem like there is no way to behave truthfully in circumstances so extraordinary or to even understand what that might mean. The first step to resolving that challenge is investigating the rules of the world you are about to inhabit before you begin to imagine the role itself. What are the edges of this map?

In Part 2 you drew a box around your personal Action Palette Diagram to try to name some of the factors based on group identities that impact your physical actions and how you play them. Each role you play exists inside a world with rules that may or may not be different from the real world. This means that all the factors you drew around your personal diagram become just one portion of what impacts this imaginary world. Let's take a look at the rules of the world for this script and see how it defines the outer box that will contain all of your physical actions in this role.

Rules of the World

Start by answering the following questions:

1. *Does the world include magical events?* Flying? Supernatural events? What are the limits of those? This could be something as simple as the lights flickering once when a character thinks of a dead relative.
2. *How does time operate?* Does it jump around? Do the characters experience it chronologically? When in history does this take place? Even if it says "Present," is it *your* present or was the author's "present" a long time ago?
3. *Are the locations realistic places?* A house? The head of a pin?
4. *Is there a shared history?* Does this world imagine a history that didn't actually happen? A change in events from our world?
5. *Do people use language in a way fundamentally different from our world?* Is the text poetic or heightened? Does it include invented words?
6. *What do you or others say about the group identities of your role?* Do you know the gender identity? Race? Sexual orientation? Are they assumed, implied, or stated?
7. *Are the physical forms of the people the same?* Are some characters animals? Space aliens?

Once you've answered these questions, do the following:

1. *Draw a large box on a page.* An example of a finished version is shown on the next page.
2. *Label each side, using these categories: Time/Place/Space, Text, Group Identities, Production.*
3. *On each of those sides, write a few words using the answers from the questions above.* By "space" I mean how do the rules of science operate. This is where you can note any unrealistic or supernatural factors of the world. You may not be able to respond about "production" until later in the process. Often actors come to understand those decisions when they see designs, as rehearsals progress, or when they arrive on set.
4. *On the "Group Identities" side, make note of the identities you know for the role you're playing.* Sometimes this raises questions about the distance between you as an actor and how the role is described. Maybe all you have now for that side are some questions to write down.

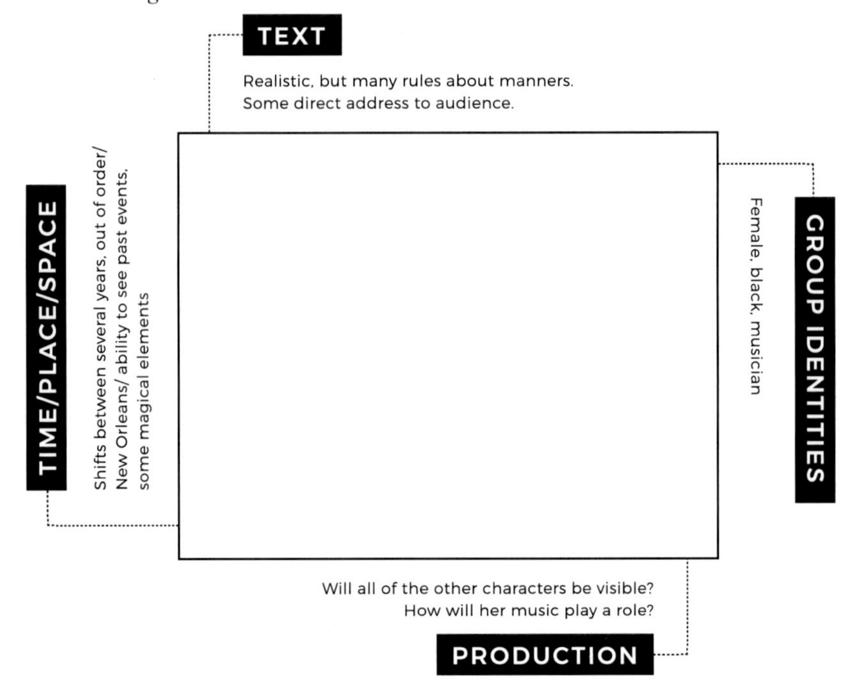

Rules of the World

These factors: Time/Place/Space, Text, Group Identities, and Production are categories that exist in every world you help create and can impact the way you play all your physical actions. Sometimes there is very little to think about for some of these categories. If the world is quite contemporary, there are no unrealistic events, the text is contemporary and conversational, and the role shares all or most of your group identities then the box this creates might not require much focus from you in the process. Still, it is important to ask these questions so you don't *assume* this to be true. The circumstances of the world of the play are not always familiar to you in your life. If you grew up in a wealthy family and the people in the play are impoverished, that is not an unreal world despite the unfamiliarity to you. Still, it may impact the way the character can play actions. We'll further investigate those distances from you personally later. Any of the sides of this box can alter the ways the world fundamentally operates for your role.

This exercise often produces a list of questions you need to investigate during your rehearsal process and with your director. That final side, Production, is probably blank right now. That's okay. Even projects where the script seems fundamentally realistic might be performed in ways that are not. Sometimes actors struggle with how a director and designers choose to shape the world and to make sense of why, for instance, a scene where they keep saying "sword," includes them holding a gun while they say it. That's why I include production as one side of this box. Every production includes decisions. Sometimes they align closely with the world the text describes. Sometimes, intentionally, they do not. Your job as an actor is to play your actions within all the sides of that box. Part of the journey in rehearsal is discovering how to make all of those sides work together.

Walk the Path

Now that you've read the script a few times thinking about the totality of the story, you can focus on the way your role interacts with and lives inside that world and those events. It's important now to engage the rest of your body. This isn't about being anti-intellectual or anti-research. It's because a physical engagement with the text early on can help these processes. If you begin to physicalize the fictional environment from the start, you bring the text closer to the lived version and improve your understanding. You also begin to build connections in your mind between the events in the text and their physical form. The earlier you start this in your process, the greater the number of connections you can build over time.[3]

Walking the Role

1. *Take the script and find a space (it can be your home or a studio) and begin to "walk the role."* By this I mean sit and stand and move to different rooms or spaces or whatever else the author tells you happens or you understand must happen based on the text.
2. *Do the behaviors and activities described in the script*. Remember, these are things like sewing or chopping wood, or cracking your knuckles, or whatever else is stated. You don't need the real objects. It is fine to use approximations or imaginary objects, especially if the objects would be dangerous. You don't need to know about the set or the blocking or anything else. You're simply imagining a physical world for these scenes and events and following early physical impulses about when you want to move or sit or do some activity.
3. *Say your lines out loud as you do this.*
4. *Interact with imaginary partners*. If the script says you hug someone, embody that activity. If you serve a meal, go ahead, and set a table or put a platter out. Use what you have readily available to approximate.
5. *Keep your journal for this role nearby and write any physical actions (transitive verbs) as they occur to you*. Remember, don't just write the activities you do, write the actions you think might be behind the impulse.

6 *You don't have to be right. Don't go line-by-line.* These actions are early hunches about what you are doing with verbal or nonverbal communication.
7 *Try it without speech.* Try this same experiment again, this time without speaking the text. Just follow the flow of behavior and activity and see if any nameable physical actions come to mind as you work.

What discoveries did you make about the way this person inhabits the world by walking the role? Does it all take place in one room? Are they always outside? Are they ever alone? How much activity do they do according to the text? How did you handle unrealistic locations or events? Did they invite unfamiliar but exciting physical actions? What was different when you tried it without speaking? Because most of us are so verbal and because words are so much a part of our expressive life, find pleasure in exploring how much gets communicated without them. However, this doesn't mean you should "mime" all the gestures that go with speech. It also doesn't mean you should mouth the words but not put voice to them. This is the difference between saying "Go to hell!" and slowly closing a book while giving someone side-eye in a way that communicates that same impulse. Both can have the same physical action. Even though you carry the script and explore the moments and events, allow verbal expression to take a back seat for a while.

Playing like this early on helps you recognize and build the role with embodiment at the center of the process. Did walking the role change your ideas about the character's wants and needs? Did it help you discover new events or realize that something you believed was a major event actually is not? Don't be afraid to let the discoveries from later experiments alter your decisions in previous ones. This work is not a straight line or a one-way street. You're digging in and getting your hands dirty. Discoveries that change your mind don't make your previous thoughts wrong. They simply make them previous.

Find True North

Now that you've read the play a number of times, twice embodying the role in some fashion, let's step back and ask some questions you can use to dig deeper. The text and events in the play are the information you have, but they often reveal other factors that impact how you play the role and the physical actions available to you in this world. I like to start with "What is my through-action?" You might be more familiar with versions such as: "What do I want?" "What is my objective?" "What is my super task?" However you like to phrase these, you're ready to ask the core questions about why your character exists in this world. What are they are trying to accomplish through their interactions with everyone else? While you may change your mind as you learn more and work on the role, it's good to begin with what you think is happening in this early moment in order to do the work that follows. Be sure to use the notes you've made about rules of the world, major events, your first read, and walking the role to help answer these questions and generate some of your own.

Five Big Questions

1. *If you got exactly what you wanted at the end, what would the world look like?* Maybe it is as simple as "They marry me." It could be something more seemingly impossible like "My father comes back from the dead." Find the north star you will aim all of your physical actions toward. *Make sure it is something that requires another person.* It shouldn't be "I understand the meaning of loss." What does the *other person* do and say if you succeed?
2. *When you look at this goal and the major events, what is your initial impression of the through-action for your character?* Try to state this as a simple version of the goal above. Using those examples, it might be something like "I capture them" or "I resurrect him."
3. *What non-text events significantly impact this person?* This could be a childhood event, something that happens in the next room right before a scene, an accident, war, etc. Try to base it on the text and not merely invent something exciting.
4. *What might be the major circles for this person's Action Palette Diagram?* Are they family, work, and partner? Are there others? Maybe one circle just for a specific person?
5. *How do the rules of their world make it harder to get what they want?* Their time and place, the words available, identities, and other factors all impact how they play actions to achieve their ultimate goal.

Your answers to these five questions don't need to be definitive. At this early stage they point you in a direction and generate more questions to explore in the work ahead. I encourage you to keep a few pages in your journal for this role and title them "The Notebook of Things I Don't Know." Use it to generate and track questions you can bring into the rehearsal process, to come back to in later experiments, and to remind you that you are in a playful process of exploration rather than trying to get things right.

These early steps focusing on the world you'll inhabit are important. Without a clear idea of the story you are telling and how it is being told, you risk making your work too self-focused and can easily slip into the comfortable habits and actions you already know. Each world a writer creates defines the edges of the map and frames the audience experience. This must mold and shape both your explorations and the choices you ultimately make. With this deepening understanding of the world, you are ready to explore more deeply the point of view and experience of this role.

Notes

1 Mueller, Pam A. and Daniel M. Oppenheimer. "The Pen Is Mightier than the Keyboard: Advantages of Longhand over Laptop Note Taking." *Psychological Science* 25, no. 6 (June 2014): 1159–1168. DOI: 10.1177/0956797614524581.
2 Benedetti, Jean. *Stanislavski and the Actor*. New York: Routledge, 1998, p. 105.
3 Lutterbie, John. *Towards a General Theory of Acting*, Cognitive Studies in Literature and Performance. New York: Palgrave Macmillan, 2011, pp. 98–99. DOI: 10.1057/9780230119468.

9 Making Text Physical

It's You, But Not You

In this chapter we'll take many of the ideas that you explored in Part 2 looking at yourself and apply them to a role. Your experience using these tools to examine your own life substantially improves the specificity and skill you bring to this next step. Your goal is to build a solid understanding of the role, how this person interacts with the world and others, and to craft a deeply investigated set of offers to begin rehearsals. I say "offers" because it is vital you don't think of this process as landing on a set of answers. You cannot know what your fellow actors, designers, and director will bring to the table. What you arrive with must be a flexible and available sense of the role that allows for experimentation, discovery, play, and change while helping you stay engaged with your partners and focused on the goals of this person.

I hear actors say things like "I haven't figured out this character," or "I've almost got this character." After the self-evaluation you did in Part 2 you recognize that your own character shifts from environment to environment, setting to setting, and group to group. Remember the concept of code-switching. Are you two different people? No. You are a single person for whom the switching itself is part of your character. As an actor that can be confusing and it's easy to worry that an audience will be disoriented by the changes. I think the opposite is true. The consistency of character that is so often cultivated by actors allows audiences to lean back and predict what comes next. The excitement of change draws us forward.

Part of the pleasure of acting is making shifts away from yourself. How is this role like you? How is it not? To help our experiments with those questions we'll employ a term used by Rick Kemp in his book *Embodied Acting*. [1] Think of this role as a "temporary situational self." This is a good reminder that the character your audience experiences is inevitably partly you and partly not you. Some of who you are as a person is a part of what the audience experiences as the character. Some of your own experiences and memories and associations may even link up well with events from the script. On the other hand, there are also certainly experiences and events with which you simply do not have a personal relationship and must imaginatively engage.

Your explorations in this chapter will help build connections and memories related to the role and strengthen them through embodied explorations. This imaginative work is more like remembering your own history than you might realize. Imagination and memory are related processes and not as separate as we once assumed.[2] As you imaginatively build these connections, memories, and craft the action palettes a clear picture of this situational self emerges. Once you craft a situational self from the information you have about the role, you can look at the distances between you and the role. This gives you an important road map for conversations with collaborators about how the production plans to bridge those distances or how you can bridge them on your own.

Embody the Major Events

You already identified the major events in the script on your second read. Select a few of those that involve your role for this next experiment. If you aren't a part of those events, select two or three scenes that feel like the most significant moments in the script for you. Maybe you witness or overhear a massive fight between two other characters that changes the course of the story. Maybe someone tells you about it later. Perhaps the first time you know something shifted is when you discover a car is gone from the driveway. What is that moment like? Major events probably impact you, even if you aren't at the heart of the event itself.

Find a space in your home or a studio to explore these events. You're going to do a version of the Physical Playback experiment from Part 2. As a reminder, this experiment is an opportunity to visit a key moment and try to understand the physical actions you might be playing. Treat this imaginary circumstance in much the same way. Even though you may not have met your scene partner or know the blocking yet, you can still engage in an imaginary version of the event that will help you understand your physical actions and begin to build an action palette. It is important that you don't imagine a set or audience or other theatrical factors. Imagine the event as it might really happen in the life of this person. There is plenty of time to worry about the constraints of performance. Let those go for now.

Physical Playback – Major Events

1. *Select an event from the script you want to investigate and find the minute or two that feel like the most important part of the event.*
2. *First, walk that event like you did in* Chapter 8 *during the Walking the Role experiment.* Speak any text. Feel free to alter what you first tried. Learn from that experiment and build on it here.
3. Do that a few times. It is okay if it changes as you imagine and reimagine the event.

4. *Find a single line/word/sentence/silent moment from this event to explore.* Try to make it the most essential moment of this event. This is the part that flashes through your mind when you remember something from the past.
5. *Put your body in the shape of that moment as you imagine it.* Are you standing or sitting? How are you turned? Are limbs crossed or holding any objects? Are you walking? What are you wearing? Where is the other person? Are you looking at them? If not, what are your eyes focused on?
6. *Now playback just this moment.* What is the verbal and nonverbal embodiment of the event?
7. *Do it several times as a loop.* If you seem to discover new elements, incorporate them with the repetitions. It could just take a few seconds or maybe it is a slightly longer moment.
8. *Make sure that you reinvest in the event each time.* Don't allow the repetition to become hollow.
9. *If it is hard to catch it all at full speed, slow it down.*
10. *After a number of repetitions, try to name the physical action.* Do any words you speak offer clues? Do the position of your body in the room, your gestures, or other non-verbal elements help identify the impulse? Remember, it doesn't have to be right. It is your best assessment.
11. *Write this down in your journal for the role where you wrote the physical actions discovered during your earlier reads.* Be sure to note the event, the other characters, and the text involved.

Repeat this exercise with a number of major events from the script. It is helpful, when possible, that all the events don't involve the same partners. Finding events where you interact with a variety of people helps later and ensures that you aren't taking too limited a view of how this person interacts in various settings. Take a few small and seemingly uneventful moments for the role and explore the physical actions for those, as well. If you only explore high stakes events for a role you gain a distorted view of how they operate. As you know from your own life, the physical actions that come out in extreme circumstances are related to, but not always the same as, the palette you use all the time. Getting robbed and making a sandwich can both happen to the same person but, probably, don't involve the same set of actions.

You may find that a few key moments of the text are enough to begin identifying the Primary Action Palette. In other cases, the complexity or the distance from your own experiences might demand the exploration of many more. There isn't an exact number. Remember that the origins of your explorations and your work should be the pleasure of voracious curiosity. Use the experiments to discover information and answer questions. Enjoy the chance to crawl around inside the world and play on your own. Explore the moments that are elusive or confusing, or that demand information and context you don't have yet.

The Edges of Expression

Once you've done that work, look for two moments to explore more deeply. Select one moment where the role expresses their physical action in a very large and visibly released way and one moment when their physical action, because of the circumstances, is only allowed a very narrow range of outward expression. This might mean that they don't even have a line to speak or the words don't come anywhere near expressing (or express the opposite of) their action.

It is important to look at these two extremes. It's easy to imagine that expression exists in a narrow band. In life, and in many scripts, there is a range of expression available to each person. Perhaps they rarely yell, and yet at one moment they do. That physical action must still be a part of their available palette. Is it played often but expressed in other ways? Is it rarely available to them? You create greater consistency in the impulses for your role if you find the extremes of their expression and look to see how they might exist in the same person. Now take these two moments and let's experiment with expanding and contracting the physical action like we did for your own actions in Part 2.

Expand and Contract – Text Extremes

1. *Select one of these two actions.*
2. *Once again, walk that event like you did in* Chapter 8 *during the Walking the Role experiment*. Speak any text. Feel free to alter what you first tried. Learn from that experiment and build on it here.
3. *Refine it through repetition to a single line/word/sentence/silent moment from this event*. Try to make it the most essential moment of this event.
4. *Repeat that over and over so it becomes a short loop of the same physical action.*
5. *Begin to expand the size and scope of the physical form*. If it is a lean forward, could it become standing up? Could the impulse start from somewhere else in your body? Where could it begin that gives the impulse greater power?
6. *Let this expansion open new ways to physicalize it*. Maybe as the original gesture expands it makes you want to move around the room or twist or use your body in ways you never would in everyday life.
7. *As you keep looping the physical action and expanding the non-verbal elements, let the verbal elements shift, too*. Release the

action/impulse from the rules of the world where you can. Do you say more, or do the words change? Perhaps you just end up making non-word sounds. Sometimes it helps to say the action verb itself to support your exploration. It is okay here to leave behind the text as written. For this experiment, you can speak/express something that is more directly connected to a released version of the action. Sometimes it helps to say what they *wish* they could say but aren't allowed to.

8. *Keep expanding this loop until you engage your entire body (including your voice) in the strongest possible version of this physical action.* It is okay to keep changing and exploring with each loop. Embody the impulse as completely and unselfconsciously as possible.
9. *Keep increasing how important or high-stakes the action is.* Even if the moment in the script isn't like that, allow this to become the singular moment you can achieve this goal.
10. *Once you have reached the most extreme version you can find, immediately snap back to the original version.* All the rules of the world that shape and squeeze that impulse return. Try to rediscover that verbal and non-verbal expression.
11. *Do that original version a few more times.* Can you feel the strength of that bigger impulse wanting to come through? How effective does this version feel now? Does it still get across the action you intend it to?

Repeat this exercise with the other moment you selected. How is it different depending on how much expression and release is allowed within the circumstances themselves? At the end of the exercise, where in your body do you feel the desire to express the action is most suppressed or held? Where does the ability to express it still feel most free and released? Sometimes doing this reveals that the text allows for a fuller expression than you recognized, or that more of the work must be contained. It may be worth repeating this pairing of expanded and restricted expression for your interaction with different characters if the rules change in very significant ways depending on who you are with moment-to-moment. That leaping hug might be possible with their best friend but not with the plumber who fixed their burst pipe for free – no matter how similar the impulse.

Embodying Unseen Events

We're watching this story because significant events occur that alter the lives of these people. Part of your job is to help the audience believe that this person continues to live a life beyond what they witness on screen or the stage. This means gleaning from the text or inventing events from their life leading up to the moment we meet them and, in some cases, events that happen during the story but are not witnessed by the audience.

If you're reading this book, you might be someone with a flair for the dramatic. That's wonderful, but do resist making this role's personal history interesting or riveting or terrible or glamorous without the support of the text. If the writer never provides any indication of a traumatic past or a horrible secret or untold adventure, then let's assume it doesn't exist. What is the value to the story or the audience for injecting that when no evidence supports it? Be content finding great joy in the regularness of their life and the difference from your own. Why wouldn't they be worthy of the attention of our audience? Everyone's life is full of events that made them who they are without those events destroying them and forcing them to rise from the fire like a phoenix only to arrive on stage reborn.

Take a few minutes to write some key events in the life of this role. I encourage you to create a kind of timeline. Space out the events on the page (s) as they are in life. If several events cluster over a few weeks or years, write them quite close together. If something else happened many years ago, leave a gap in time on the page. It helps to realize that their life probably had long stretches of relative calm that formed them, too. Start with the events (if any) that are described in the text. Maybe you or another character recall a shared moment from the past. Perhaps you keep referring to "that one night seven years ago" when clearly something happened but nobody says what. Or maybe there is a character we hear the name of often who never appears. What might have happened with them?

Once you establish the key events from the past that are supported by the text, begin to imagine life during the gaps. Were there other notable events? Did they move around as a child? Did they experience loss or violence? Were there great successes or joys? Did they fight in a war? What are some events that helped shape the way they live in the world and the actions they play? A few are fine. Add these to the timeline.

Looking at this history, select a couple of events that cover a range of experience and ages. Perhaps they occur in different places or with different groups of people (family, peers, etc.). With these few events selected, you'll now explore a physical playback of these events. It is worth saying before you begin that this kind of exploration can be fun, and it can also be difficult. Some of the events might be disturbing or have produced trauma in this person. Be careful with yourself without being precious or overly fragile. If you're interested in the job of acting, you're excited by imaginative engagement in some of the most extreme moments in the lives of people and how they respond. That said, because we know that memory and imagination are linked closely in the brain, imagining a difficult moment might feel like a kind of experiencing of it, though it is not the same.

Maybe you have also undergone difficult or damaging experiences in your own life. Imagining events that are similar or produce similar responses might bring up painful memories of those events, too. Still, part of your work as an actor is developing a set of skills to explore and replicate these dramatic experiences in a character's life with a level of control that allows you to maintain your own sense of self. It may help you to have something nearby that reminds you of a pleasurable experience in your life and helps ground you in the present. Maybe you can plan to move to another room or do an activity quite unrelated to the role immediately after the exercise. I don't worry that you'll forget who you are. It's just that it may help you to release the imaginative engagement and return to a calmer mental state after the experiment. We'll start from a wider and longer exploration of the event you imagine occurred, find your way to the core event and, ultimately, the physical action.

Physical Playback – Non-Text Events

1. *Select an imagined event from your character's past you want to investigate.*
2. *Begin by walking about five key minutes of the event.* Like you did earlier, speak aloud to imagined partners and "hear" their response. As difficult as it might be, physically engage in the event. Because there is no text for these moments, you'll simply improvise as you go.
3. *Obviously don't do anything where you (or others) would be physically hurt.* In those cases, use imaginary objects or close your eyes and imagine those moments as specifically as possible, then continue physicalizing when it would be safe to do so again.
4. *At the end, take a few moments and note several physical actions you played.* Are they similar to ones from the text? Are they brand new? Remember not to simply describe what you did physically. What were the actions that generated the impulse? If you punched someone, perhaps the action "mangles" or "stuns" captures the desire under it, for instance.
5. *Repeat this playback of the imagined event again, this time only exploring four minutes of the event.* What is an even more core/essential version? Can you start later? End sooner? Don't hesitate to change and discover new imagined elements of what you say (or don't say) or how the event plays out physically.
6. *Continue this process a number of times playing back a shorter and shorter version of the event. Each time make notes about the actions.* Perhaps the same ones keep coming up. Maybe the event keeps changing.
7. *Keep working to find the core moment – the instant that flashes through the mind of the character when they think of this event in*

their life. Eventually you should have shortened it to a single word/sentence/silent moment to explore.

8 *Put your body in the shape of that moment as you imagine it.*

9 *Now playback this moment.* What is the verbal and non-verbal embodiment of the event?

10 *Do it several times on a loop.* If you seem to discover new elements, incorporate them with the repetitions.

11 *If it is hard to catch it all at full speed, slow it down.*

12 *After a number of repetitions playing the moment back, see what verb fits best.* Do the words you speak offer clues? Do the position of your body in the room, your gestures, or other non-verbal elements help identify the impulse? Remember, it doesn't have to be right. It is your best assessment of the core action for this event from their life.

13 *Write this down in your Physical Action Journal for the role.*

As you work on these key events, it is worth noticing the patterns that emerge. Do they all involve the same people? Are they all in the same room or city? Do they all begin or end the same way? Sometimes patterns reveal valuable information about the role. Other times they are a sign that you should select different moments to explore or expand your own imaginative engagement to enrich the view of their past. If you need to, take breaks between these explorations. Step out of the room, allow your breathing to deepen, or do something that brings you fully back to yourself. You might even do this between moments of great joy for a character. Actors must be robust and able to move in and out of challenging imaginative circumstances, but part of building that skill means finding what helps you step away and knowing when you need to do that.

Primary Action Palette

You've spent a good deal of time in the embodied world of this role. You walked the entire journey through the text a couple of times. You explored key events in the text and expanded and contracted the desire and expression of physical actions. You investigated their past and the embodied events that might have contributed to who they are and how they play actions in the present. Take a look now at your journal for the role and all the pages where you made notes about the physical actions you discovered and explored. It is time to craft a first draft of your Primary Action Palette. Much like the one you did in Part 2, use these collected observations to step back and get a picture of their essential character. It is a starting point. Just like with you, these are the actions they are primed to play most readily and in the greatest variety of circumstances.

Generating the Primary Action Palette

Reviewing your journal for this role:

1. *Read all of the physical actions you wrote a few times to see what patterns emerge.*
2. *Identify actions that appear multiple times.* If there are no repeats, are there some actions that seem to describe a category they use a lot?
3. *Make a list of about 10–12 of those actions to create the Primary Action Palette.* If you are still struggling, look for ones that you think others would use to describe this person.
4. *Be sure there is a selection that represents them in various social situations in life* (home, work/school, with strangers, specific other characters, or other groups that are a major part of their regular interactions).
5. *Read it aloud.* Does that sound like them as you understand them at this early stage? Do you have an opinion or judgement about the person that palette describes? Does anyone from your own life comes to mind when you say those actions aloud?
6. *If it seems entirely unlike them, look back on your list and switch in some actions that offer a more complete or accurate view.* Be careful not to exclude or eliminate actions just because you don't like them or feel judgemental about the way you might be viewed playing them.

Think of this Primary Action Palette for the role as a beginning rather than an end. You'll discover so much in rehearsal and when working with partners that you cannot imagine right now. Having your own thoroughly explored ideas fuels shared discoveries and helps your partners and director expand and shift how this person fits into the world you create together.

Take a moment to look at the palette you created and think about the categories of actions you selected. What is the balance of primary actions (stabs, hugs, etc.), conceptual actions (adores, rejects, etc.), and metaphorical actions (fuels, shrinks, snows, etc.)? Is this balance very much like your own action palette from Part 2? If so, is that the right fit for this role? It might be helpful to think about the way this person interacts with others and shift the balance of the palette to support those impulses. For instance, if all of their language and behavior is direct, simple, and unfiltered, maybe a palette that is heavier in primary actions will help you find the embodied way those impulses live. On the other hand, someone with incredibly florid language and lots of hidden intentions might also use mostly primary actions, creating a sharp and biting undertone to their treatment of others. There is no right answer, but these choices help you understand and craft how this person thinks about the world and what they need from others.

Some of the actions in their palette might also be ones you play regularly or are even ones in your own Primary Action Palette. Part of exploring this temporary situational self is finding where they do and do not overlap with you from everyday life. If you're cast in this role, it is likely there are things about you the director felt were a good fit. It is okay for your palette to overlap with theirs. As you continue to explore, ask yourself if those actions come up in similar situations or different ones. Perhaps they play those actions differently than you do. You might also discover later that the rules of the world force those same actions to be expressed in different ways.

How Their Puzzle Pieces Fit

In this next experiment, let's return to the idea of personal fronts from Chapter 6 to see what else you can learn about this role and the embodied way the character moves through life. We'll start to understand how these factors might shape the actions they play. Remember that personal fronts are the *versions* of self that people perform in different settings. I'm me, whether I'm giving a lecture or making silly faces with my nieces. People are always a *version* of themselves, and yet certain actions are excluded or added so the personal front fits the rules they perceive for that setting.

Sometimes the reasons for these shifts are about status. Every relationship and every setting have status dynamics. Some are obvious, like a boss and an employee. Others are murkier, like between partners, or friends, or in families. Status can also shift over the course of a script. As you begin to consider this factor below, be careful about taking on too modern a version of this idea. Can you investigate how status works in this particular world and with the rules as you understand them? Is it shaped by the larger social dynamics of the time and place? Sometimes the character will have their own view of their status that doesn't fit modern sensibilities. Find the pleasure of playing actions within those restrictions. They still have needs they want met and ways to make that happen.

Personal Fronts for Settings

1 *What are the different settings this person operates in?* This might just mean locations they visit in the script, but it could also mean environments we never see but know exist. For instance, perhaps we only see this person at home, but they speak a great deal about their job. At the same time, even different rooms in the same home could be very different settings depending on your role.
2 *Take a sheet of paper and create a section for each of these settings.* Let's assume there are at least two, but it is probable there are three or more. Unless there is a particularly special situation, keep it to four as a maximum. That likely covers the vast majority

of interactions they have and the key elements that make up the way they present to the world.

3. *In each of these sections of the paper write the setting and a word or two to describe the role they play.* For instance, "kitchen/head cook" and "dining room/servant."
4. *Along with this, write the status level they have in that setting using a scale of 1 (lowest status) to 10 (highest status).* In the example above, this same person may be a 9 or 10 in the kitchen but a 2 or 3 in the dining room. That will become helpful not just in the physical actions they can play but also how they are able to play them.
5. *From your earlier notes about physical actions from reading and embodying the text, write the actions that appear in each setting.*
6. *Also take the actions from the Primary Action Palette and write them in each section.* Some actions may be written in many/all sections. Others might only appear in one setting.
7. *Think about their job or hobbies and the activities that go with them.* A cook likely does a lot of stirring, chopping, seasoning, and baking. It can be a pleasure to see if some of those job-based activities can become some of the verbs you use for the Primary Action Palette. Do "stirs," "chops," "fries," or "boils" seem like actions that fit their behavior in the script in other ways. Add those in to the sections where they apply.

Once you've done this, take a look at these two to four settings and the list of actions that help form the personal front this person presents in each. Which ones would the audience witness most? Are there ones that we never see on stage? How much overlap is there between the physical actions you wrote in each section? Are there fronts where they make a massive shift? You may even notice that two adjacent scenes in the text require a major shift in personal front. That can be an exciting experience for the audience. What information does the audience get from watching the palette of physical actions shift so strongly and suddenly? Does that tell us something about this person's point of view about others and their status in these settings?

Of course, someone can be standing in the same location and, depending who walks in or out, there can also be a sudden shift in the available physical actions. Much like you observed in your own life, this person has actions they reserve for certain relationships. If your parent walks out and your partner walks in, your palette changes. Some people you interact with are simply covered by the palette for a particular setting. Other relationships are unique and require their own list. Take a look now at the key relationships this person has in the script. These are ones where there are significant interactions or a great deal of history before the script begins. Let's repeat this task with these relationships to better understand how they alter behavior no matter the setting.

Generating Relationship Palettes

1 *On a sheet of paper, create a section for each of the major relationships.*
2 *Write the name of the other person and the title (if there is one) for this relationship.* For instance, "Remi/lover" or "Bryn/teacher."
3 *If it is clear to you, write the status differential between your role and this other character using a scale of 1 to 10.* For instance, maybe you are a 5 and they are a 7. Or perhaps it is 10 and 2. This will require more exploration in rehearsal, but often the script offers plenty of information to estimate and explore how status impacts the available actions. Remember to resist imposing your own views on the relationship. How do *they* experience the status?
4 *From your action journal for the role and the lists you created for the Personal Fronts for Settings exercise, write several actions that you play with this other person.* Are there actions that you only play with them? Are there categories of actions (primary, conceptual, or metaphorical) that you lean on in this relationship?
5 *If you're struggling to find some, read some scenes between you and them.* What do the events and dialogue in the text tell you that might help name some actions?
6 *For these key relationships, name five to ten actions that solidly fit.*

Once you have created these lists, step back and look for the differences. Are there elements of the relationship that seem to be missing? What physical actions can you add to make the picture more complete? If there are major changes in the relationship during the script, what actions seem to capture the relationship most broadly over the entire arc? How much is status impacting your choices? Don't forget, a low status doesn't mean you can only play seemingly low-status actions, but it may mean that certain strong actions are forced to be masked or shaped in surprising ways.

Organizing the World

So, you have all of these lists, categories, actions, and settings, but how on earth will you ever keep those in mind when you're standing there with a partner saying your lines? Not to worry. Let's take a moment to create a unified picture of all you have done so far. Just as you generated an Action Palette Diagram for yourself in Part 2, we'll do the same now for this role. All of this work primes you to be present to play from well-informed and well-investigated impulses in the rehearsal room. For this next experiment, use the sample diagram after the instructions as you go. The visual will help make them clearer.

Action Palette Diagram

1 *Gather together the Primary Action Palette, Personal Fronts for Settings, and Relationship Palettes for this role.*
2 *Using a blank sheet of paper, write all of the actions from the Primary Action Palette in a cluster in the center of the page.* Use pencil, as you'll have to move and rewrite words a number of times.
3 *Select a total of 3–4 relationships and/or settings that cover the majority of this person's identity.* If a setting is essentially covered by a specific relationship, there is no need to also make a circle for that setting. Better to focus on people than places.
4 *Using the action lists for those settings and/or relationships, write several key actions from each list in clusters around the actions you've already written down.*
5 *Look for the physical actions, if any, that occur in all of these lists and make sure they're at the very center of the page.* They become the core of the Primary Action Palette.
6 *Organize them so that you can draw overlapping circles as you did for your own Action Palette Diagram in* Part 2. Cluster the actions for each setting or relationship together and write the actions that exist in multiple palettes toward the edge of the cluster so they can be in the space the circles overlap.
7 *Move words around or change the way the circle overlap as you work.* Once you're done you can always redraw it.
8 *Label the circles with the name of the other character or the setting it references. Also write your status for each relationship (e.g. if you are a 6 and they are a 10, write "6/10").*

When you're done, take a moment to scan this diagram and see if it leaves out any major element of how this person interacts with others over the course of the script. Are there major settings you've left out? Is there a relationship you kept out that now seems more important and should be added back? Could each of the circles use another physical action or two to make sure they are clearly distinct from the other parts of the diagram?

The sample on the following page is what a finished diagram might look like. For the example, I imagined a role where we see the person with their partner, Devon, a best friend named Kai (who they have feelings for), some co-workers, and customers at a retail store. You'll notice that not every circle needs to overlap with all the others. Take a look at the actions they play for both Devon and Kai versus either alone. The similarities and differences in the palettes provide some interesting dynamics if all three characters were ever together. It may even reveal why the relationship with Devon isn't working.

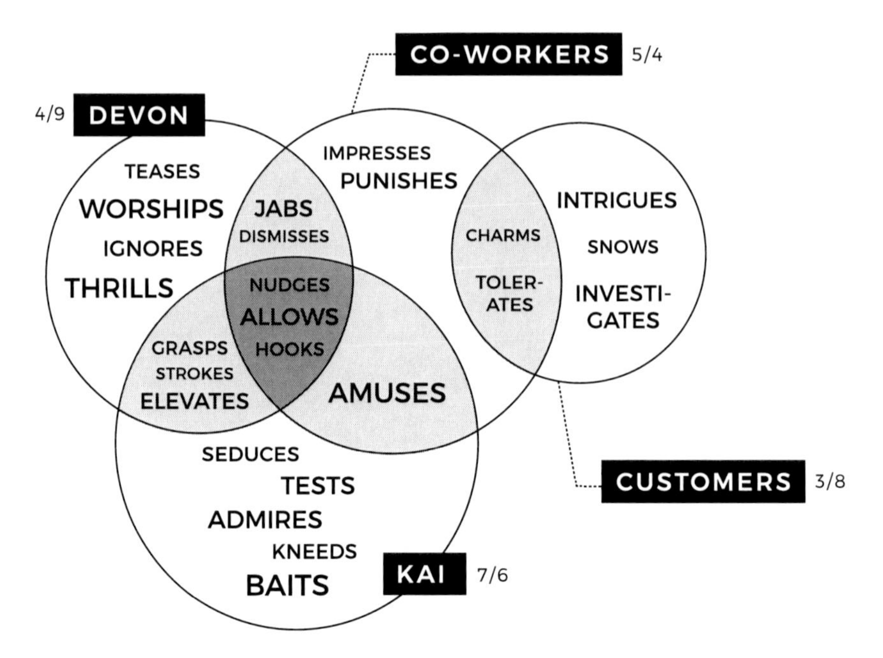

Action Palette Diagram for a Role

The Action Palette Diagram helps you organize all of your thinking about physical actions for this role so far. The diagram doesn't contain every action you've imagined from all your experiments, but it contains the primary palette you created and a good sample from the major relationships and settings from their life. Just as you observed about yourself, these are key habits they return to over and over and with variation in different settings.

Looking at the Action Palette Diagram you created, you begin to get a picture of the character of this person as the audience experiences them. These actions are tangible ways you will explore playing in rehearsal to understand how this person works moment-to-moment to achieve their larger goals in the story. It is important to remember that thinking about the role in these embodied ways, using verbs as a core vocabulary, helps you craft a sense of the embodied version of them. They keep you focused on *doing* something to your partner and require you to keep your attention on how it impacts them. It keeps your focus outside of yourself. You're beginning to think about their physical impulses and, likely, how they differ from yours. Let's experiment with the Action Palette Diagram you created. Keep it close by as you work.

Changes Tell Stories

Something happens when a new character enters a room. How do they change the dynamics? What new possibilities arise? What paths close? Using your diagram, you can now see that if you're in certain settings or with certain people and someone else in the diagram arrives – a new set of physical actions becomes available to play with that person. In the same way, when someone exits a scene, the actions you wrote only in their circle become unavailable to play. By reserving certain actions for certain settings or people you help the audience to experience a clear shift and to understand the distinct differences between relationships. Imagine the person you're playing has two sisters. Making certain actions distinct for each of them rather than simply having a palette for "sisters" allows for greater specificity and clarity in your playing. It also tells the audience important information about the history you have with each of them. If, during a conversation about where to have lunch, you play "brushes," "resists," and "placates" with one sister while the other gets "babies," "steadies," and "scolds," the audience receives a lot of valuable information and gets curious about the history. Experiment now with some of these shifts by embodying moments of change.

Switching Actions

1 *Select two actions, one from your Primary Action Palette, and one that is only used in a relationship and does not overlap with other circles.*
2 *Begin by physically exploring the primary action.* It is fine to start small. What gesture or impulse feels related to this verb?
3 *Eventually let the action actively include the use of your entire body as you play.* Find a version in the same way as in the Expand and Contract experiments. It may help to slow it down, especially for actions that are quick or short.
4 *Once you find this, let the character for whom you picked the other action begin to enter your mind.* You don't have to know who the other actor is. Your imagined version is fine. Begin to imagine them as you explore this action.
5 *Does that change the way you embody it?* Remember, it is far beyond what would be considered "realistic" physical behavior. Does it give the action new purpose or potency when you have a clear target? How do you play with action in *this* relationship, not just generally?
6 *Let the importance of this relationship and any status difference strengthen the action and determine the shape.* How do you imagine them/their body responding?
7 *Keeping your body fully engaged, begin to shift to the action you selected for this relationship.* Don't do it all at once. Let the shift happen over time as your body explores the expanded physical form of the verb.

8 *Really imagine them as you play this action. Let it become clearer as you repeat and refine the physical form.* How do you imagine them now? Does this action fit them better? Do you imagine a response? Does a line from the script come to mind?
9 *Explore saying the name of the character or a line from the text that seems to fit this physical action.* How does speaking aloud change the way the rest of your body experiences the action? Does it help make the imagined partner clearer?

Just exploring a physical action in the abstract is rarely as clear or helpful as having a specific target and partner. Sometimes doing this reveals that the action you selected isn't quite right. You may find that you want to rename either the choice from the Primary Action Palette or the Relationship Palette once you explore it in an embodied way. You tried a single action in this exercise, but I encourage you to do this for each of the key relationships from the text. It is important to see that the same action manifests in different ways depending on the target. *The key purpose of playing an action is to change the other person.* That means staying curious and attentive to how it lands and what will most impact this specific partner.

Let's take this one step further. Sometimes a person enters a room and they are from another setting where you usually play very different actions. This offers an excitingly complex moment. Do you only play the actions that fall in the overlapped area of the diagram? Can you play actions with one person while the other witnesses them or do they need to be private? Think of the awkward circumstance of a parent walking in on you while you're alone with a romantic partner. How do you choose the actions that are possible in this suddenly blended space?

Switching Palettes

1 *Select two characters you interact with but who have little to no overlap on your Action Palette Diagram.*
2 *Pick one physical action for each of them to explore. Pick a spot for each of them in the room and take a moment to clearly imagine each of them in their spots.* It might help to close your eyes and let the image of them grow and solidify.
3 *Explore and expand the action you first tried in the previous experiment.* Once again, find an embodied version of this action. Perhaps it is different this time. Are you already imagining a person/target?
4 *Now, actively imagine one of these two other characters as you play this action from your Primary Action Palette.* Discover how

it alters the action. Really aim the action at the spot in the room you imagined them.

5. *As you did before, shift over a number of seconds from this action to the one that is only for this relationship.* How does the shift first begin? Where in your body? Remember to explore saying the character's name or a line that comes to mind that might fit the action. Don't forget status. Keep the image of them and their response alive and strong as you work.
6. *Once you have a fully expanded the embodied version of the action and are saying a line or the person's name, begin to imagine the other character as clearly as possible while continuing to do this action.*
7. *Continue doing this while you turn toward the place in the room you selected for them.*
8. *How, in your body, do you know this action feels wrong for them?* Does it "break the rules" in some way? What happens if you said their name as you did this action? Does your body change the way the action is played almost against your will? Do you laugh? Feel sick at the thought?
9. *Keep doing it with a clearer and clearer vision of this person until it feels unsustainable.*
10. *When that happens, shift quickly to the action you selected for them instead.* How does it feel once again to have the action and the imagined partner match each other? How do you use your body differently when imagining this other partner?
11. *Keep going until you feel firmly in your relationship with this second character.*

Many people discover just how firm the walls are between certain relationships when they attempt to play an action that is mismatched. In our own lives, and certainly for many roles, the palettes are clearly defined and distinct. There are many theatrical circumstances, however, where dramatic events rely on playing actions the audience will see as "wrong" for the relationship. Whether it is abuse, or transgressive sexual relationships, or simply betrayal and lying, it's vital that you see not just where *you* draw lines but, importantly, how the person you're playing draws lines differently from you.

You can explore in this way many times, using different combinations of characters to understand these overlapping circles. Sometimes you'll discover that a physical action should move to an overlapping area of the diagram because it makes perfect sense in the other relationship. You might discover that you need stronger differences in your choices to allow for clearer lines between relationships. All of this, of course, will also shift as you meet your collaborators and explore with a director how they imagine you as a part of the whole world being crafted.

Huge Changes Over Time

The early work of reading a script, embodying the text, and defining your initial take on the actions, palettes, and diagram is work that you can do for almost any substantial role. Some texts are simpler than others, but most well-written roles interact with a variety of other characters and we see a complex enough imagined world that you can define several settings and relationships.

Sometimes it helps to go beyond this work before rehearsals begin because the role requires a massive transformation over the course of the script. Perhaps a major discovery occurs. Maybe they experience a psychological or physical trauma. If you're working on a film or a series, there may, over time, be major shifts in a role that you must track, even though scenes or entire episodes might be filmed entirely out of order. Whatever the cause, a single Action Palette Diagram might be too straightforward to carry you through the arc of the role. In these cases, it can help to explore the key shifts in available actions from before to after these events.

Before and After

1 *On a number of small slips of paper, write each of the actions from your Primary Action Palette for this role. On two other pieces write "before" and "after."*
2 *Place the "before" sheet on one side of the room and the "after" sheet on the other.*
3 *On a final piece of paper write "the event." Place this in the middle of the room.*
4 *Take the slips of paper with actions written on them and place them in the room in relation to where you imagine they are used the most.* If they are used all through the script, place them in the middle of the room.
5 *Select 2–3 actions that are not part of the existing list that are only possible before this major event. Write them down on pieces of paper and place them over by "before."*
6 *Do the same thing for actions that only occur "after."*
7 *Take some time to explore and embody the actions from "before."* While you can use the whole room, try to stay closer to that side of the space. Don't assume that everything was wonderful or terrible before this event. Who are these actions played on? People we never meet in the text? Someone who dies during the story? Someone with whom the relationship transforms? Expand and contract their form as needed and desired. Make sure you imagine playing them on a partner, not just alone or disconnected.
8 *Explore these actions without the knowledge of the event.* Free yourself of knowing what will happen. Do not embed the future event in these actions.

9. *As you explore them, begin moving toward the paper that says "the event."*
10. *When you arrive at that spot take a large inhale and exhale of breath.* Let sound come out on the exhale. It doesn't have to be words or text or anything recognizable. Think of this as that key moment of the event. Is it the moment of marriage, or loss, or discovery, or release?
11. *Let your body spend a few moments discovering and exploring some physical form of this event.* It doesn't need to be a literal playback as we've done before. Let it be an outward expression of the internal sensations of the event. If it is one of the events you explored, perhaps play that single action you discovered that captures the key moment.
12. *After you explore the event itself, take a step toward "after" and explore the actions from the after section.*
13. *Keep moving further and further from the place in the room where the event occurred as you explore only the "after" actions.* How are they different? Where can you feel the residue of the event itself? Does the partner/target of your actions change? Or does just their response to your actions change? Is there someone you want to play actions on that you no longer can? Why are those "before" actions no longer possible?
14. *Finally arrive at the piece of paper where you wrote "after." Do those actions just a few more times.* What does it feel like to have some distance from the event? Does it change those same actions now that time passed?

This works best when there is a clear and definable moment of shift in the text. Did you discover anything about the event itself by exploring in this way? Maybe it happens offstage. If so, what physical actions might be part of the event? Some actors find it helpful when they are offstage/camera to play some of the actions from the event itself to embody the experience before re-entering after the unseen event. If the event is particularly emotional or traumatizing for the character, remember to take some space to distance yourself after working. Perhaps there is an object, or location, or song that helps. It is important to check back in from time to time about the distance between this situational self and your everyday self. Being an actor means being excited by exploring and investigating the extreme moments of human experience, but it is okay and important to let go of the imaginary circumstances when you're done.

If there is not a singular event or moment where the shift occurs, it can still be useful to do this experiment and think of the event as a kind of time jump. Imagine that another character hadn't seen you for a long time. What differences would they notice? You can also simply do this for the beginning and end of a particular script. What key physical actions are available at the beginning that are gone by the end? What actions are not possible at the beginning but become possible because of events in the script.

As I mentioned above, before-and-after experiments are also especially valuable for working in film, television, and other media where your performance is recorded in a non-linear way. Taking time to define physical actions that are possible before and after key events helps you imagine the embodied arc of the performance and offers a strong, simple reference to review right before you shoot a moment and keep the continuity of their journey.

The Beckett

As Samuel Beckett famously wrote, "Ever tried. Ever failed. No matter. Try again. Fail again. Fail better."[3] Most of your actions in most moments of performance fail. This isn't your fault, per se. If your actions succeeded all the time it would make for very little conflict and a fairly short script. You'll come to notice times when you seem to be hammering away at the same action over and over for minutes or pages with very little change. In these instances, it's important to investigate the moment-to-moment differences more deeply. If you play the same physical action for that long, audiences grow restless with the predictability of your work. Stanislavsky addressed this need for, what he called "adaptation" when first attempts fail.[4] Our minds are built to separate the important signal from the noise.[5] Too much of the same action becomes noise your audience can ignore. *The more you reward us for close attention, the more we look for the next exciting shift.*

When you find a section like this in your text, explore a set of cascading physical actions. These actions are like a ladder of attempts, one building on the next. Often the verbs share some traits, but there can also be surprises and shifts along the way. For instance, you might be asking someone for money. Perhaps the cascade of actions is "charms," "prods," "shoves," "chokes." Of course, it could go in exactly the opposite order, or you could find that the cascade looks like "prods," "shoves," "*charms*," "chokes." This shouldn't be arbitrary or based simply on what might be fun for you. It is, as always, grounded in what the text supports. You must look for both the words and behavior described as well as how other characters respond. There is no sense in crafting a cascade of physical actions that contradicts the response of your partner.

Cascading Actions

1. *Find a section of the text that seems to be a long stretch of a single physical action.*
2. *Select the action you think describes what is being played generally over the arc of the section.* Be careful to really select a transitive

verb. Remember, "flirts" is not helpful because you need to add "with" to the end. Try something more like "seduces" or "wins." Build it around the specifics of this moment.

3 *Make a list of several actions (perhaps five) similar to the one you selected with varying degrees of strength/intensity.* For the example above you might use "lures," "nudges," "pulls," "warms," "tickles," and "adores."
4 *Select a few lines of the text and say the original action you selected, followed by the line of text.*
5 *As you do this, imagine the partner in front of you.*
6 *Begin to expand the embodied expression of this action as you work.* Work over several repetitions to find an expansive exploration of that action.
7 *In this expanded physical way, start to shift between this action and some of the others you found that are similar.* How are they different in their physical expression? How do they alter the way you speak the text? Does one seem to feed the text more than another?
8 *Keep working like this, shifting between these similar actions and bits of the text to mix and match.* You may work in a methodical way, matching different actions one after the other with the same text, or you may find it more helpful to let the physical expression inspire a line of text.
9 *Eventually, see if you can explore a few specific orders for the actions with the text.* What order feels like it builds alongside the text? Are you continuing to imagine your partner as the receiver? How are you imagining their response to these actions?

When you complete this experiment, make a note in your text with that order for the cascade of physical actions. In rehearsal you may find there are even more, or perhaps the order or list will shift when you have a real partner. For now, you have an embodied starting point from which to offer your partner and director some clear and specific way to build this set of moments.

This is also a valuable tool for monologues. Too often in both performances and auditions, actors wash entire sections or an entire monologue with one or two physical actions. It is challenging when your partner doesn't respond verbally, and you know you won't be interrupted. Still, you must live in these moments with the possibility that they *might* respond or interrupt. You are still receiving information from their behavior, just like in life. You are still in the constant loop of action, curiosity, and attention. The shifts in a monologue must reflect the success or failure of your actions based on your partner's behavior since they do not have text. In life, we frequently adjust the length and detail of a story based on how our audience responds moment-to-moment.

That same quality of attention must be present in your work, even though the text is fixed.

If They Aren't Like Me, Who Am I?

In many cases, the experiments with the story and text in Chapter 8 and this deeper exploration of the role in Chapter 9 is sufficient work to do before rehearsal or filming begins. That said, there are many times when the role you're asked to play is very distant from you in certain ways, or the nature of the text is so far from "reality" that you struggle to understand the rules of the world and how they impact on your actions. Perhaps the most common complication is the way you do or don't share identities with the role. In the next two chapters we'll look at how physical actions might help you work under these conditions before we move on to working with partners and groups in Part 4.

Notes

1 Kemp, Rick. *Embodied Acting: What Neuroscience Tells Us about Performance*. New York: Routledge, 2012, pp. 210–211.
2 Schacter, Daniel L. and Kevin P. Madore. "Remember the Past and Imagining the Future: Identifying and Enhancing the Contribution of Episodic Memory." *Memory Studies* 9, no. 3 (2016): 245–255. DOI: 10.1177/1750698016645230.
3 Beckett, Samuel. *Worstward Ho*. New York: Grove Press, 1983.
4 Benedetti, Jean. *Stanislavski and the Actor*. New York: Routledge, 1998, p. 79.
5 Clark, Andy. *Surfing Uncertainty: Prediction, Action, and the Embodied Mind*. Oxford: Oxford University Press, 2016, p. 53. DOI: 10.1093/acprof:oso/9780190217013.001.0001.

10 Building Bridges to You

Distance from the Role

Part of the joy of acting is finding pleasure in taking on elements of a persona that are distinct from you. If you're incredibly tidy there is delight in playing an absolute slob. If you're deeply ethical you might find it quite fun to be scandalously unethical for a role. There is also, of course, the endless series of not-doctors, lawyers, and cops who "play one on TV." Few people, other than actors, get to try on these differences in life or through their work. Exploring and having an appetite for those distances can be one of the best parts of the job.

Sometimes those distances can also be challenging or confusing. This is especially true if your gender, race, nationality, physical abilities, or other factors potentially core to your identity and experience of the world are different from those of the role. These challenges also differ depending on where you are in the world as you read this book and your sense of how the audience might receive you in this performance. It is important to be honest with yourself about your experience of these distances. You might find them unimportant to you or insignificant for your work on the role. Perhaps the fact that they are written as a man and you're a woman doesn't seem relevant or confuse you much. That's fine. Others might experience this as so unfamiliar that they cannot grasp a way into the work. While we strive to tell stories that impact people in universal ways, your personal experience is singular and unique. I want to offer a few thoughts on how to acknowledge these distances (which everyone has in different measures), find pleasure in that distance when you can, and provide a few ways you might think about them through the lens of actions.

Building Bridges

Recall the list of group identities we explored for you in Part 2. That list (which is certainly not comprehensive) included age, education, socioeconomic class, religious belief, family membership, biological sex, gender identity, gender expression, sexual orientation, citizenship, race/ethnicity, skin color, and disabilities. For many actors, the canon of plays used in academic settings or working in diverse groups on "classic" plays means frequently confronting a feeling that a writer did not originally imagine someone like you in the role. In those cases, it can be important to honor and acknowledge the distances, recognize the ways you are close to the role, and explore how you might build bridges you can imaginatively cross each day as you work.

Often this requires a conversation with the director at the beginning rehearsals to understand how they perceive the world of this production. I encourage you to engage in this conversation as a part of your work. If you are Japanese-American and the role was written for someone white, it is okay to ask how the director understands this as a Japanese-American person for this production. What does that mean? What questions and complications might this raise about the time and place? I don't know how they will answer, and the answer may not satisfy your need for information. Still, directors should think critically about the bodies on stage or screen and the way the audience experiences them. As a collaborator, it is worth starting a conversation to understand how the director imagines the audience will experience you.

The experiment below is not about finding a way to alter or erase who you are and replace it with other identities. You are who you are and shouldn't need to pretend that isn't true. At the same time, your work as an actor always requires imaginatively stepping into someone else's shoes. Often, we talk about that in terms of circumstances, but identity influences the way we respond to circumstances. We all have personal and cultural histories that are deeply rooted. My goal here is to give you a way to use the tool of physical actioning to explore how the identities of the role influence the way that person navigates the circumstances of the text so you can find versions of those actions from your unique perspective.

Building Bridges

1. *Write down all of the group identities you know for this role on individual pieces of paper.* The Rules of the World work you did in Chapter 8 will help with this task.
2. *Don't only write the ones that are different from you.* If it is unspoken, but assumed or implied, also write that. For instance, if it is clear from the text that, as written, the role is not a person with a disability and that is also true of you, make note of that.
3. *With you at the center, place the pieces of paper around you in a circle. Put them as close or as far from you as you sense them to be.* If you are from Russia and the character is from Russia, you can keep that one right with you in the middle. If you are from Uruguay and the character is from Russia, put it as far away as *you experience* that idea. It is important that it is personal for you. Find the distance and scale that feels right.

4 *Once you've done this for each group identity, slowly turn in a circle and look at the physical map you've created.*
5 *Focus first on the elements that feel close to you.* Notice the ways you and the role overlap or are in close proximity.
6 *Now explore a physical action that feels like it fits that element of your own identity.* Perhaps something from your Primary Action Palette from Part 2. Don't worry about being literal or being able to explain why it feels connected. If you are from the United States and so is the character, find an action that instinctively feels like it expresses your relationship to being from there. As an example, maybe "shrinks" does that for you. Perhaps "confronts" is a better fit. It is just a touchstone for the similarity between you and the role.
7 *Experiment with expanding and contracting the action.* In each step, it is good to explore in this way to understand all the places in your body where you are able to express this action.
8 *Does this role experience that identity in the same way?* Maybe your physical action for this identity works well for the role, too. Check to see if the circumstances make the identity feel different to *them*.
9 *If necessary, select a different action for their version of this shared identity.* Sometimes the time, place, and events mean that even overlapping identities benefit from different actions.
10 *Now turn your attention to the items that feel like they fall somewhere in the middle distance from you.* For example, perhaps you feel a small sense of distance from the education level of this role. Find an element of identity where you and the character are somewhat different from one another.
11 *Make this a two-step process. First, explore an action that feels related to your own sense of this element of identity.* Using the example of education level, this is an action you play on a teacher, or one related to a specific memory of education for you.
12 *Now, select a different action for the role in relation to this element of identity. What is their experience of it?* Maybe you have a moment from the text that makes that clear. Perhaps a line or behavior where their relationship to it is expressed. What was the physical action for that moment?
13 *Once you find that, begin to explore moving back and forth between them. Travel from the center of the circle toward that spot in the room and back as you work. Feel the sense of that movement/action in your own body.* How is it different from your own sense of this element of your identity? How can you feel a clearer sense of ownership over this less familiar action? Is this an action you feel connected to in some other way? Might that help you understand their relationship to this identity? Can you find

some pleasure in the action they're able to play and the difference from your own action?

14. *Now, turn your attention to the group identities with the greatest distances from you.* Often these are the ones that we feel most strongly about ourselves and we think others see when they look at us. They are frequently visible physical traits like skin color, race, gender, and disability that can profoundly impact the way others treat us.
15. *Begin to work in the same way here. What is an action that comes to mind when you explore your own sense of this element of identity?* It may be hard to capture in a single action. Is there a word, or phrase, or experience? When and how do you feel most proud of this identity? It doesn't have to be the case, but it may help to begin from your own sense of self in a positive way rather than starting by thinking about distances.
16. *Now think about this same identity element for the character. Is there a moment from the text where you think that their sense of this identity is present?* Is this something they contemplate in the world of the play? Sometimes in texts these identities are not overtly explored by the writer because they were assumed. Still, you may be better able to recognize how their identities operate in that world because of your distance from them.
17. *Begin to explore a physical action from this moment or their palette that best represents their relationship to this part of their identity.* See if you can avoid working off of opinion or judgement and focus on their experience of it as best you can assess.
18. *Once you've found that, explore back and forth between these two actions, the one for your identity and the one for theirs. Cross the space back and forth as you switch actions.* You may discover that you need a bridging action between the two because of how profoundly you experience this distance. Sometimes that distance also generates an emotional response.
19. *As you move back and forth between actions and in the space, find a physical action, or maybe two, that become(s) a bridge from one to the other.* For instance, if your action for your group identity is "astonishes" and the one for their identity is "flicks," perhaps you build a bridge with "dismisses" and "swats" to move between the two.
20. *You may also find, despite your sense of the distance between you and this role, that they actually have quite a similar relationship to their identity as you do to yours.* Their privilege based on that identity might be quite different, but the way they operate within it might be similar.

Once you've done this for a number of key identities for the role, explore using these bridges as a tool at the beginning of rehearsals or even as a warm-up. You might try reversing the direction at the end of a rehearsal or performance to acknowledge a return to self and the work of bridging those differences. Remember, the shift in these actions is not about taking on their identity or erasing your own. The goal is to give you space to understand how this person's identities operate, as written, and understand the actions that will help you tell the story. It offers you permission to stand in that action as yourself, despite the potentially unfamiliar nature of how they operate in relation to identity on the page.

One important note. If you're operating from a place of privilege and being asked to make a shift to a place of less privilege, you must start with the assumption that it is a long bridge to build. Imagine, for instance, that you share racial, religious, and ethnic identities with a role, but you grew up a citizen of the United States and they are living in the United States without citizenship documents. Don't underestimate the significant privilege you likely hold in comparison. There is a necessary humility in that.

We started the book with the importance of finding pleasure in curiosity. Honestly assessing the gaps in your experience and knowledge so you can fill them with that voracious curiosity is part of your work. To pretend they don't exist undermines that core value. The result of this experiment is likely to be a series of questions to explore in rehearsal. If you are from Saudi Arabia and the role is from the United Arab Emirates, you may know a lot about that country and how relevant identities operate, but maybe not. That not-knowing is an important part of your journey of curiosity. Let that distance generate questions you must answer and research you must do to better understand they operate inside the text. Actors get the great gift of learning so much about people and the world project after project. Use these gaps in your own knowledge as a call to dive in and fill them with experience and information.

Sometimes a major concern is imagining yourself in a world where the research and realities of that time and place wouldn't allow someone who looks like you to do some of the things you do or say. This is especially true in modern productions of older or "classic" plays written in Russia, Europe, or the United States. It is even more common in academic settings. Perhaps you're Haitian and being asked to play a banker in the 1940s in Atlanta married to a woman played by a white actress. There are, inevitably, difficult questions about interracial marriage in the United States at that time that the play may not address. These circumstances are deeply individual to each actor and each project.

My greatest encouragement to you is to start that conversation with your director early, especially if they do not start it with you themselves. If you need help working out your place in the world of this piece, then ask for partners in that journey. There may be clear ways to make it make sense. In other cases, the director might very much want the audience to grapple with the difficulty of what you would have faced in that time and place. Unfortunately, it is also possible the director is simply not interested in exploring this with you or might not understand the way it is impacting your ability to work. In each case, more information is better. The sooner you understand the way your presence and identities intersect with the production the sooner you can understand the bridges you need to build for yourself and in conversation with collaborators.

The Audience Experience

One further challenge in performing a role with different identities than you is anxiety about audience perception and response. Will they accept me in this role? Will someone say something to me after the performance? Will they understand that I am that other character's mother even though we don't look alike? Sometimes these worries come from real previous negative experiences. Other times actors have positive experiences that reinforce their ability to be seen and accepted by audiences in these ways. Many actors also get positive responses from audience members who share their identities and gain the opportunity to see themselves represented in the work through the actor's performance. Remember, though you experience the audience collectively, they are also uniquely individual in their experiences.

There may be some comfort to you from neuroscience in these sometimes thorny circumstances. Audience members cannot help but see your performance as a blend of you and the character. That is true for every actor in every role no matter how many of your group identities overlap. The character, in most cases, does not actually exist and never has. What the audience has is all the words and behaviors of that imagined person embodied as you. That "conceptual blending" of character and actor is an inevitable part of viewing you in the role.[1] It requires real resistance for an audience member to keep those two elements separate the entire time they watch. Whatever your journey of seeing yourself in the world, the journey of the audience is often far simpler. You are that person to them in the most essential way. Audience members who resist this blending or cannot help but comment on their difficulty seeing you in the role are engaged in an active resistance to the experience of watching performance. I find that revealing about them and no reflection on your work.

I don't believe that simply doing this one experiment, Building Bridges, will resolve these questions for you. What it can do is offer you a tangible way to do something inside the work that helps you operate in rehearsal, have some agency in the exploration, and better grasp the way this character plays actions in the world the writer imagined. Not all of the bridges will be hard to build, and in those less fraught circumstances, it will help you find clarity and pleasure in the opportunity to inhabit someone unlike you in various ways.

Note

1 Fauconnier, Gilles and Mark Turner. *The Way We Think: Conceptual Blending and the Mind's Hidden Complexities*. New York: Basic Books, 2002, pp. 266–267.

11 Special Circumstances

The majority of your work as an actor will probably require playing a single role in more or less realistic circumstances in front of a live audience and/or a camera. As technology, the industry, and theatrical forms shift, however, more of the job includes elements that don't fit the model that actor training and many acting methods envisaged. There can be a lot of delight in the variety of ways you get asked to represent humanness in the course of a career. This chapter covers a number of common circumstances actors face and offers experiments to help you apply the tools of physical actioning. It is important that you feel confident using the same values and skills in your work even when the nature of the performance feels unfamiliar. It will ground you and give you confidence that you understand the fundamental purpose of your work.

Working with Multiple Roles

Playing multiple roles is the most common circumstance you're likely to face. The risk, of course, is that one of the roles feels more exciting or interesting or important to you and the others receive short shrift. It's important that you do the basic work to investigate every role, however briefly they appear. In some cases, this means much more invention on your part because the text offers so little. Remember, avoid making excitingly theatrical choices simply to keep yourself interested if there is no support for them in the story. If you're playing a soldier with one line, there's very little use imagining that character as a secret spy for the opposing army who has a career infiltrating enemy lines. How could you possibly offer that up in your brief appearance? What would that add to the audience's experience? Better to think about where that soldier most likely came from. What does it mean to them to deliver this news or speak up in this space? What is their status? How long have they been travelling or how far from home might they be? These things might influence this single moment and could be supported by the relatively small nature of this person in the text.

When you are working on multiple roles it is vital you generate a Primary Action Palette for each one. Even five physical actions that are core to how this person operates and how others experience their character is a valuable foundation. When you rehearse or perform, *find an action from that palette to embody right before you enter as that person*. This is particularly helpful if you're switching quickly between roles. Perhaps you have a costume change and are running around right up to the moment of entrance. Even a tiny pause to *think* that physical action and engage your muscles in some way as you are about to enter clarifies this shift. It can be as simple as finding a moment to exhale, think the action verb, and make a fist or shift your pelvis or anything that feels part of the embodiment of that action.

This is most helpful if, as you rehearse, you explore that palette each time before you work on that role. Linking the embodied physical actions and the role over and over may help link them in your mind and spark those connections when you enter those scenes.[1] Select an action for right before you enter that is significantly different from the other roles, particularly the one you played immediately before this moment. This does not mean that you have to make some kind of massive transformational physical change for each of the roles. The nature and scale of those shifts are between you and the director. It does mean that you can make even the subtlest shifts in impulse and the actions available to this person – a shift that makes sense for acknowledging the differences between any two people.

Scaling Your Performance

No two rooms are the same and your performance is altered by your distance from the audience, partners, and a variety of other factors. Perhaps the three greatest impacts on the nature of how you embody a role are size of space, rules of the world, and medium of performance. Live performance spans a wide range of venue sizes. You could be performing across a table from a single audience member or on a football field in front of tens of thousands. No matter what the circumstances, you must not retreat from your core moment-to-moment job – changing your partner by playing physical actions.

To grapple with the shift in scale, it's helpful to return to the idea of expansion and contraction. By now you've done many experiments where you found a more "realistic" version of an action and expanded it out to include a more and more visible embodied form. Filling a larger or smaller space is not simply a function of how much of your body you include in the action or about the size of your gestures. It is also about the way your external expression contains the strength of the internal desire. You can change the scale of your performance through the way that internal need and external expression exist in harmony or contrast. It is good to do this next experiment in a larger space. Somewhere you can move around and run if you want and not be worried about making noise or knocking into someone or something.

Dials of Engagement

1 *Begin to walk around the space. As you do, find a pace that feels like it is about the middle between you standing totally still and you running at full speed.*
2 *Let's call this pace "5." "0" would be total stillness. "10" would be total chaos. You're right in the middle.*
3 *As you walk around at this pace, use your "internal sense" to explore what is going on inside your body.* This can be difficult to understand at first. Perhaps this is a sense of your heart rate. Or maybe how quickly your mind seems to be thinking. Or perhaps it is related to your sense of your breath. Maybe it is a harder to describe sense of urgency or anxiety or tiredness. Ideally, it is some combination of these factors and more that generate a sense of the "energy" inside you. Do your best, even if this is an unfamiliar idea.
4 *Can you put a number to that internal speed? Is it also a 5? Is it a little faster or slower in your experience than what is taking place on the outside?* For this, think of the scale as "0" being total stillness (like a deep meditation) and "10" being total chaos (heart racing and breathing heavily as if you've been running for a long time). An internal 10 is a feeling of energy racing around inside you or lots of adrenaline from being surprised/excited/afraid/etc.
5 *Think of each of these external and internal sensations as "dials." You have an external dial that describes the visible movement of your body through space and an internal dial that describes the internal sensations of movement, thought, and energy traveling through you.* For most people, these dials are close to the same most of the time.
6 *As you walk around the space, begin to imagine that internal dial turning up a notch from 5 to 6.* Experiment with how to generate a sense that things on the inside are moving just a bit faster than things on the outside. Maybe start with a shift in how you use your eyes to look around the room. Perhaps a change in your breathing. Some people find that imagining a scenario helps to generate a faster sense of movement within them. Can you do it without allowing significant visible change on the outside? Does it require imagining or visualizing that internal motion?
7 *Now turn the external dial down to 4.* Remember, this is just a bit slower. Make sure that includes everything. Your arms swing slower, your feet step more slowly, even turning your head to look at something is a bit slower. It is like the air becomes thicker around you and slows you down.
8 *Feel this distance now. Your external dial at 4. Your internal dial at 6. Where in your body is that most noticeable?* Are there places

inside where you don't have a real sense of the movement? Can you imagine the way your blood pumps to those parts or imagine the nervous system sending electrical impulses there?

9. *Now turn the internal dial up to 7.* How can you help increase that sensation? Where does it begin? Does it require some visualization or circumstance you imagine?
10. *Turn the external dial down to 3. Now the distance between these two dials begins to feel quite noticeable.* Perhaps it is like a rubber band being pulled in opposite directions. Try imagining a pot of water on a stove. The internal dial is the water heating up and moving more and more quickly. The external dial is the pot itself – the lid firmly on containing that internal motion.
11. *Turn the internal dial up to 8 now. Just 2 steps away from total chaos.* Can you feel a sense of that energy and need pushing against the barrier of your skin? Does it leak out? Where? A tension in your hands? Your eyes? Your breath?
12. *Turn the external dial down to 2. You should be moving very slowly now.* Can your whole body be engaged? It is almost like a form of slow motion. All the while there is this sense of internal movement driving and pushing to get out as the container of your skin maintains this much slower speed.
13. *Turn the internal up to 9.*
14. *Turn the external dial down to 1*, moving as slowly as you can without being still.
15. *Turn the internal to 10. Things churning and racing now.* Really try to imagine it reaching every single inch of your body from the inside. All the way to fingertips. The back of your neck. Behind your knees. Is it difficult to sustain? Do circumstances or images come to mind?
16. *In a moment you're going to go ahead and turn that external dial all the way up to 10.* You're going to let it match the inside. You'll take the lid off the pot and let the dials match again.
17. *Feel what that anticipation does to the internal sensation.*
18. *Ready? And GO!* What are all the ways you allow the speed and energy and size of what was being held inside release through the body? Do you make a sound, laugh, or scream? Do you leap? Do you thrash or pound your chest or run or drop to the floor? How long does the release last? Is it over in second? Where and how does it linger? Do you want it to last longer? Why?
19. *When you feel like things find a new equilibrium, what numbers do you land on?* Perhaps you're right back to 5 and 5. Maybe it settles into something further apart than when you began.
20. *When you finish, make a few notes about the experience.*

This experiment generates an incredibly wide variety of responses in actors. You may experience that moment of release as great joy. Others think of a panic attack and a desire to flee the room. Some have an emotional response. Others feel like the sensation disappears immediately. Still others simply cannot generate that internal sensation. The extreme disparity of the internal 10 and external 1 is an extreme version of something we all experience at moments. There are times of excitement or danger or anxiety or joy where what is allowed to be expressed on the outside does not match the sensation we have of size or speed or need on the inside. Some people live in a state like that.

This tool might be helpful for many roles in understanding how a person navigates the tension between their strong desires and the rules of the world. It also helps you explore how to scale to different sized performance venues. Filling a larger space will not necessarily mean larger gestures or bigger engagement. It may simply mean turning up the internal dial on the actions. That sense of additional need and urgency might be enough to help the audience experience your actions over distance. If not, you can explore turning up that external dial as well. It probably won't be external 10 and internal 10 – that would be hard to sustain and exhausting to watch for a long time – but an external 6 and an internal 7 might help give you the image needed to keep your performance communicating past the usual distance. Keep adjusting so you find the balance that is truthful for those circumstances but communicative over the right distance.

Experiment now with some specific physical actions for the role in this way. How does the expression change if the need is stronger but the external expression must remain contained? What about the opposite? What is it like if there is very little internal need but an expansive physical expression of the action?

Dials of Engagement – Actions

1. *Select a verb from the Primary Action Palette for the role and a line of text where that action might be clearly expressed.*
2. *Say the line of text infused with the action. Let your body engage in whatever way makes sense.*
3. *Think of this as both dials being at a 5.* You are able to express the internal desire and action through a simple and released physical form.
4. *Start to experiment with turning the internal dial up and the external dial down.*
5. *Go step by step.*
6. *Discover how this lower external number contains the expression, including language.* It doesn't have to be about speed of movement. Which parts of your body become less available? Which words disappear? Are you forced to change the shape of your body or use of self to help contain the action? Does an activity help you contain it?

7 *Don't lose the growing internal need and rising number on that dial.* Keep the water heating up inside the pot. Let the desire to play that action push against the constraints.
8 *Continue playing like this, turning the dials up and down to discover how it impacts your expression.* Notice when circumstances or scenarios come to mind. What would explain the inability to release the action? Does it have to do with status or relationship? The rules of the world?
9 *Don't forget to turn the dials the opposite way, or experiment with both being turned up or turned down.* Do personalities come to mind? Are there people you know who seem to have certain combinations of dials?

Experiment with several physical actions from the Primary Action Palette or the Action Palette Diagram. Beyond simply discovering how to scale the performance to the circumstances of the production, you may also discover that this person exists in a state of internal/external imbalance. Does establishing their general state for the dials help explain their Primary Action Palette? Are there moments when the difference is most pronounced? Does the balance of the dials change when they are in a scene with a specific other character? Maybe you've experienced that speeding internal dial when someone you love entered a room, but you weren't able to show it on the outside, or during a moment of great fear.

Working On-Camera

Many excellent books already exist about acting for the camera specifically. Some of the challenge of this adjustment for actors is about scaling their performance. In many respects, working on camera is working with a very close audience. Close enough that, with some shots, they see every little thing happening in your eyes or with your breath. Exploring the Dials of Engagement can help you adjust to the camera. It is possible that the technical demands of a shot will require you to keep very still. That should not prevent you from playing actions on your partner. In some respects, that simply means you must turn the external dial way down and play with the internal dial to ensure the action is coming though easily and simply with your gaze, words, or breath. Maybe your external dial is 3 and your internal dial is 6. Perhaps the camera is so close that 2 and 4 are more than enough to communicate your action. This adjustment may need to change shot-by-shot, but your core task remains the same.

We are always balancing the strength of our impulses and how much we allow them to manifest in a visible and audible way. There are many factors that impact on how much permission we have to allow others to see our impulses. You might experiment with many variations of internal and external dial numbers and record several takes of a moment or scene on your own. What internal number is so low that you cannot see the action manifested visibly when you play it back? What is so strong that it overpowers the audience because of the closeness of the camera? The more you see your work in this medium as existing on a continuum rather than a completely different job, the more you can use the skills and vocabulary you already have to support the work.

It is also important to remember that, just like a play, there are rules for the world of a film or series. The tone and speed and way of working likely fit in broad categories of the genre as well as quite specific rules for the world and aesthetics of that project. Don't forget that the box around your Action Palette Diagram – the Rules of the World – are an important part of understanding *how* to shape actions to make them visible to the audience. The medium is a part of that.

Un-Real Rules of the World

This same idea applies to projects where the rules of the world are significantly different from our own. Often people have too narrow an idea of what constitutes realistic physical behavior. Think of images of people grieving or fleeing danger or winning a game. Perhaps you've seen people have ecstatic spiritual experiences like speaking in tongues. One time, while working for a cleaning company, I ran from a house after discovering it was infested with fleas. My behavior was realistic, but not very contained. There are plenty of moments in "real life" when our physical engagement goes far beyond simple gesturing. Even the most naturalistic text might demand more and bigger visible physicalization than you imagine. That said, there are also scripts where the writer actively breaks the rules of our physical world.

Sometimes the unfamiliarity of this distracts actors from the *purpose* of the shifted rules. The writer is almost certainly attempting to explore something fundamentally human by asking you to behave in ways unfamiliar to humans. Rather than seeing this shift as daunting, embrace it as a kind of liberation. As you already explored with yourself in Part 2 and with a role earlier in Part 3, we often have strong impulses to action that must be resisted or squeezed or formed in order to become acceptable and succeed given the circumstances. The Dials of Engagement experiment above explores that tension, too. It can be particularly thrilling, if unfamiliar, to work on a project where those options expand. Why retreat from the chance to escape something we grapple with every day?

Some scripts include rules of behavior that are much more limited than our own. Perhaps that is because of the time period, or clothing, or rules about status, or simply the nature of the writer's particular aesthetic. These also have the potential to produce exciting challenges for your exploration of physical actions. In what unique ways does this person manage to navigate the restrictions of the world and still play actions and get what they want? Might there be certain actions that are especially effective inside these restrictions? Many actors find that the tighter the restrictions the more large, expansive actions become possible. There can be tremendous joy playing "eviscerates" by simply stirring your tea.

Musical theatre also requires a kind of performance where there the rules of the world are different from our own. With some exceptions, most people don't earnestly break out into song in day-to-day life. I encourage you to ask many of the same questions you would for any other role. You are still playing actions during the song. Perhaps the embodiment is more expanded, but no less directed at a partner (which might include the audience) and no less full of need to accomplish something specific.

My main admonition is to resist focusing on the unfamiliarity of the behavior described in a script. You risk abandoning the need to play strong actions on your partner. *The codes of behavior for any world are simply rules that govern how the physical actions can be played.* Your job is to fill that behavior with strong, clear, and specific actions aimed at your partner so it doesn't turn into merely an empty shell of inflection and manners or so broadly played that we cannot see the human in it in any meaningful way.

Auditions

So much of being an actor is not acting, at least not the way we usually think of it. You spend many days preparing auditions. People throw around statistics like "Successful actors get the job one out of every hundred auditions or more." I've spent a good deal of time coaching actors for auditions. It isn't unusual for them to have only a day or two with the material. Often that material is made up of tiny bits of a much larger script, and sometimes they have no access to the rest of the text.

If you have access to the full script, by all means read it. That provides the opportunity to do some of the work in Chapter 8 and Chapter 9 as a preparation for the audition. If not, I suggest reading another script by the same writer or, if it is media you can watch, another episode of that show or a film by the same writer/director. None of these are the same as the preparation you'd have for the role itself, but there are some important ways you can explore physical actioning in this context. Here are a few simple questions worth asking, if you don't have the entire script, to keep you thinking with the vocabulary of physical actioning.

For media where you can watch another episode:

- What are the rules of this world?
- Are there actions I see being played by many characters?
- How expanded or contracted does the expression of physical actions seem in general?

For films where you can see other examples of the writer's or director's work:

- Does the writing feel similar to the samples I have to work with?
- If so, can I ask the same questions above?

Once you gather all the information you can about the project, look at the text you have and begin to generate a Primary Action Palette. You may not have a lot to go on. It helps to simply identify the actions played in the script segments you have. At the very least, select the first physical action you play. It provides an important and clear task when you go into the room and a way to start your audition with a strong choice focused on your partner. For on-camera auditions, I encourage people to find actions for key moments if they are listening for a long time without responding. As the camera watches you listen, keep curiosity and attention alive to stay engaged and drive to the next moment when you can play a physical action on text.

More and more often actors put themselves "on tape" and send in recordings, even for projects that they will ultimately perform live on stage. There is good and bad about this, of course, but you must seize the opportunity to explore a variety of actions and record multiple takes. Watch your work to see if your actions seem to land on the partner. Are the internal and external dials calibrated right for being recorded in this way? Might another action communicate more clearly in this environment? Don't waste time imagining what they want. All you can do is gather as much information as possible about the rules of this world, use that information to craft an action palette that serves the script and lives within those rules, and then play those actions as fully and clearly as possible in the audition. Make clear and specific choices and commit to them fully.

Non-Humans

There may be moments when you play non-humans. I suppose I mean other animals, but I also mean robots, artificial intelligence beings, and even objects. I played Saturn in first grade, for instance. Though, admittedly, I did not select any physical actions for the role. It seems worth mentioning this challenge mostly because so often we talk about the work as reflecting humanity and helping the audience see something about "what it means to be human." You could, understandably, struggle to see your role as a chicken, for instance, in these descriptions. That said, assume the job is the same. For whatever reason, the writer felt animals were a better and clearer way to say something to us about our humanness.

Consider generating palettes of actions that help you embrace the differences between their impulses and yours. You already did this earlier by generating Physical Action Palettes for a role and recognizing that these actions are inevitably different from your own Primary Action Palette in a variety of ways. Perhaps you should select more, or fewer, primary, conceptual, and metaphorical actions based on those distances. Maybe there are actions to avoid because they generate an emotional response in you or assume a human experience that this being does not have.

Are there actions that support the physical life of this role in some way? If you spend time observing a chicken in order to play a chicken, you might notice the abruptness of their steps and head turns and noises. Perhaps a whole palette of actions that, when embodied, encourage that kind of abruptness like "pokes," "flicks," "revs," and "dodges," or even something as obvious as "pecks." This kind of exploration helps merge the physical demands of the non-human role with a goal-based set of actions that you can play with the text and through all the physical life of the work. On the other hand, if making this role non-human is simply a metaphor, perhaps there isn't so far to go. Don't get hung up on that distance if the script simply takes for granted that the characters think and behave much like people do.

Invisible Acting

Many actors do voiceovers, voice animated characters, and read audiobooks. While you are not seen performing these roles, your work on embodiment absolutely comes through. The simple act of selecting physical actions to play helps provide life and specificity to the recording. Often, you'll work alone in a recording space without acting partners. The more you engage an imaginative sense of playing actions on that invisible partner the clearer it comes through in your voice.

Whenever possible, stand up as you record. This helps you embody the actions, even a little bit, and infuse that into your vocal engagement. If you have time, it is worth walking the role as though it was going to be fully embodied in performance. The practice of walking the role stays with you as you go to record. The memory of those impulses helps you avoid flattening your performance or getting caught up by the sound of your own voice in the headphones. Many talented and experienced voice artists can offer far more detailed advice about working in this way, but if you do nothing else, resist thinking that all that matters is your voice. Your voice is physical.

The Reality of Other People

You came a long way in Part 3. You began by exploring how to read a text with action and embodiment in mind from the very beginning. Then you experimented with generating action palettes and, ultimately, an Action Palette Diagram to understand the role's relationship to the world. You explored how physical actions help you understand events from the life of this person outside the text. From there, you experimented with your distance from them and how actions might help build bridges from your own sense of self to theirs. Lastly, you looked at some special circumstances where using physical actions and embodiment can help overcome the unfamiliarity of the medium, the role, or the world.

Everything you've done so far you can do by yourself if necessary. These experiments don't require collaborators or partners, though certainly you can do them with a class or collaborators on a project. In many cases you can already imagine how having your actual partner with you will make the experiments even more rich. Some actors worry that this detailed work will rob them of spontaneity in rehearsal and performance. Others become frustrated that the plan they crafted so diligently seems to fall apart when faced with the reality of another person who isn't at all as they imagined. *All of this work was preparation for collaboration.* You do it so that you enter the collaborative space with a deep understanding of the text, an existing physical experience of the role, and a whole set of ideas and offers to make in the shared laboratory of the rehearsal room.

The greatest joys of the work are those moments when you and collaborators explore together. Discovering that your ideas of the role must shift or your partner inspiring a sudden change in your actions is exactly the point. Those epiphanies are made more possible and more fruitful by what you bring to the collaborative space. It is a sign that the process is working well. If you're lucky, your collaborators in theatre, film, and new media will help build and shape and uncover and shift these explorations. If not, you have a strong and rigorous set of tools ready to do your work and be an actor of integrity and insight.

Note

1 Lutterbie, John. *Towards a General Theory of Acting*, Cognitive Studies in Literature and Performance. New York: Palgrave Macmillan, 2011, pp. 98–99. DOI: 10.1057/9780230119468.

Part IV

Experiments for the Classroom & Ensembles

12 With a Partner

If you did the work in Part 3 in advance of rehearsal or shooting, you are at an excellent point to begin any collaborative process. As we've discussed, it is important that you think of it as a place to *begin*. A director, and the choices of your fellow actors, inevitably alter your understanding of the piece. It is your job to come in with strong choices for the role grounded in the Rules of the World and a thoughtful examination of the text. The work in Part 1 through Part 3 provide a way of working to do exactly that while ensuring that you embody each stage of your investigation and build your ideas with tools that smoothly shift from individual work to collaborative work. Your palettes of actions ensure that even the language you use to think about the role connects to physical impulses long before you step on set or into a rehearsal room.

Part 4 shifts to experiments you can do in pairs or groups in a classroom or ensemble setting. They build directly on your individual work and focus on playing in response to others. Ideally, you'll be able initially to explore this part in a classroom under the guidance of a teacher. Having this book also allows ensembles and directors to use these tools in rehearsal to break down some of the old ideas we have about how a script moves from page to performance. I grouped the experiments moving from the smallest collaboration (just two people) to larger groups. Unlike Part 1 through Part 3, these experiments are not necessarily a progression. Some are an excellent fit for every group or project while others become useful when you're struggling with certain elements of a role or production. The final chapter of this section is full of experiments to build specific embodied skills whether you're working on a particular text or not.

For ease and clarity, I have imagined that you're continuing to work on the same role throughout. You can certainly also use the experiments in Part 4 to work on a single scene for class or sometimes without a specific text in mind now that you have the core tools of physical actioning. Some experiments have a number of variations. These are by no means exhaustive. The variations blossomed out of discoveries in classes and the rehearsal needs of specific projects. It is thrilling to see how others take the experiments and alter them, generating new ways to play with embodiment specific to their projects. I encourage you to do the same.

Too Much to Manage

Being in the presence of another person produces a flood of information for your brain. You are hearing and processing their speech while receiving a massive amount of non-verbal information at the same time. You are engaged in mirror processes, working to predict their behavior, and having a physiological experience in your own body of what they are doing. Incredibly, you're also engaged in a variety of thoughts that have their own embodied impulses as you generate speech to reply to them. As an actor, add to that a variety of other obligations (blocking, cues, etc.). All of this happens while you attempt to keep predictive processes from blinding you to changes in your partner despite the repetition of events performance after performance.

While there isn't an easy way to lessen that cognitive load of the job, this chapter leads you through a series of experiments to help you and a partner find tools to maintain a continually embodied focus on one another in an attempt to harness what your brain is already doing. They add variations and changes that force you to work in direct response to a partner and upset your tendency to retreat to habit when you get stuck or as your ability to predict their actions calcifies due to repetition. We'll focus on living in a responsive, immediate, and embodied way with another person. After you have explored this chapter, you and your partner can also return to some of the experiments in Part 3, looking at key events and shifting palettes, and do those together.

Slow Down, See More

Let's begin by taking the time to simply be in your partner's presence. When I teach a group, I prefer to cover a large area of the floor with movement mats (like the ones you might use for tumbling). It allows students to move their bodies in more ways without fear of injury. It isn't necessary, but it does help. It also helps to wear clothes you feel free to move in without the need to frequently adjust. They shouldn't make you feel extra self-conscious, but should be form-fitting enough to allow your partner to see how you move and take in your physical actions clearly and easily. It is worth saying something about body image here. Lots of movement/workout clothes are difficult for people who struggle with body image and/or eating disorders. It is more important for this work that you can move as freely as possible and keep your focus on your partner. If in doubt, wear clothes that support your ability to do that rather than expose you in ways that make it difficult to concentrate on the work. Give yourself the best possible circumstances to put your attention where you want it to be.

Once you and your partner are in the room together, let's start with an experiment that slows down the action/reaction loop and turns your attempt to play physical actions into an imaginative tactile experience. This is similar to the "space substance" Viola Spolin describes in *Improvisation for the Theater*. [1] Michael Chekhov also speaks about "molding space" in *On the*

Technique of Acting. [2] These ideas serve the valuable function of slowing down the action, curiosity, attention loop. They provide an opportunity to observe the moment-to-moment interaction so you can experiment and make conscious choices where you usually respond habitually. We'll call this "thick space." This is a longer exercise. It helps to read it a few times before jumping in. For the experiments in Part 4, if you're not doing these as part of a class or rehearsal, consider asking someone to serve as the teacher/director and read the directions aloud while the rest of the group works. If necessary, you can also record the instructions with long gaps and play it back while you work.

Thick Space – Part 1: Finding the Space

Both partners should do Parts 1 and 2 together, though you'll only turn your attention toward one another in Part 3.

1. *Stand or lie down with your eyes closed.* You'll be here for a few minutes, so lying down is great if you can, but stand if the floor is too hard for that.
2. *Focus on areas of bare skin, maybe your face, your feet, and/or your arms. Can you feel the air moving across that skin?*
3. *Send your attention to where your skin makes contact with your clothes. Feel the clothing in contact with that skin. Can you also feel the air against those clothes?*
4. *The room you are in is filled with a substance you know well: air. It's in motion around you. Each time you breathe or move you also stir and shift and alter that substance. Try simply waving an arm back and forth. You can feel the way it pushes air around, creating a small breeze on your hand or arm.* You might also feel it as you exhale through your nose. The air passes over your lips and down the front of your body.
5. *Standing or lying with stillness again, consider that the air also provides some resistance as you move through it.* It isn't much, and you're very experienced at pushing past that resistance to walk or leap or whatever else you want to do.
6. *Now, as you stand or lie here, take a few minutes to "scan" your entire body's contact with the air. Sometimes it helps to start at your feet or head and go methodically up or down. Feel the air where you can. Where you can't feel the air, simply imagine the way it contacts your skin or clothes. Be sure to include spots you might forget.* What about the air between the small of your lower back and the floor? What about the air in your armpits?
7. *Develop this map of the way your whole body touches the space and feel how much area your body takes up in the room.*

8. *Once you have that image, make an imaginative shift. Rather than the volume of space in this room being filled with air, I'd like you to begin letting that air "thicken up" and become a different substance.*
9. *Let it happen over about a minute. Bit by bit imagine the space becoming thicker. It doesn't have to be heavy. Just something you'll be able to feel resist your movement more clearly. Something that would slow you down and require more effort to move through.* Imagine that it is in contact with your entire body – just like the air. Imagine it as something you can see, and yet can see through (a bit like water, though let it be a unique substance just for this exercise).
10. *Now pick a part of your body to begin actively interacting with this new thick space. Start small.* Maybe a few fingers on one hand. Maybe just turning your head a bit.
11. *Let moving that part of your body add to your understanding of this imaginary space.* How thick is it? Does it have a texture? A color? A smell? Let it be pleasant to explore.
12. *Make sure this imaginative space is thick enough that you must use more of your body to move through it.* Even moving an arm involves effort in your chest or a clearer sense of pushing into the floor to get the leverage to push against it. Allow it to slow you down enough so you have the time to observe all of the ways it interacts with your body and the way your body impacts this imagined substance.
13. *Bit by bit, expand that movement and allow more and more of your body to engage in exploring the space. As you move, see if you can imagine the way your movement changes the space around you.*
14. *When you extend an arm out, how does it cause ripples and waves in the space away from your body and across the room? Visualize this with each movement.* Become aware of the way every movement of your body might alter the space close to you or even all the way across the room.
15. *Once you have these ideas (A. the thick space resists your movement so you must use more of yourself, and B. each movement causes a ripple/wave in the space away from your body) begin to travel through the thick space.* Be sure to remember that even taking a step involves pushing space out from between your foot and the floor in order to make contact. Turning your head is a kind of stirring the space that causes ripples and waves, too.
16. *As you begin to travel around the room remember to play with the engagement of your pelvis, rib cage, and back – don't let it become limbs moving on a solid block of your torso.* Can your steps originate from your pelvis? Could a reach begin with an impulse in your chest or even your feet? How much of your body can interact with and engage the thick space every single moment?

Don't just walk, either. Explore being low or wide or all the movement you rarely do on a day-to-day basis. Let the thick space liberate you to play with unrealistic physical expression. How do those movements change and ripple through the space differently than walking?

17. *It doesn't need to be incredibly slow motion, but it is slower and more engaged than just moving through the air so you can observe yourself as you work.*

Now that you are imaginatively engaged in the thick space, let's use it to explore actions. Just like when you expanded and contracted actions earlier in the book, the thick space provides a way to expand the physical actions well beyond everyday activities or speech into something that engages more of your body in a visible way.

Thick Space – Part 2: Playing Physical Actions

1 *Rather than simply being in the space, begin to think of the space itself as your partner. Each time you move, you play a physical action on the space.*

2 *Start simply. Find different ways to "stir" or "slash" or "pummel" or "hug" the space.* Remember that these are primary actions. There is a clear embodied version of each.

3 *It is fine to stir with your hand and arm, but let the thick space invite you to explore a variety of ways. If your goal is to stir your partner, what ways might you use your body to achieve that most completely?* What makes the thick space respond in the strongest and clearest way to your stirring? Do that. Find that. It is okay to start with the obvious version of the primary action, but be sure to expand it and explore it in other ways.

4 *It is also okay to not always have a name for the action you're playing with the space.* Maybe you discover a way to use your body that allows for a really strong imaginative response. That's fine. You don't always need to be able to put the action into a word in the moment.

5 *The space is always receiving as a partner. Whatever you do, the space moves imaginatively in response and allows your physical action to impact it and create change.* Enjoy the receptive partner. Spend several minutes simply partnering with the space.

6 *You use your body to play an action, see how the thick space responds in your imagination, and allow that to inspire the next way you'd like to move and change the space.* Do this for several minutes or more.

Now that you have the sensation of playing physical actions in thick space, let's take the final step and shift your focus to your partner. As you do this, remember the sense of playfulness and strength you found alone with the space. When shifting to a real partner, people often retreat from this embodied exploration into a more everyday use of their body or simply gesture to make actions clear. Avoid making storytelling or your partner "getting it" a priority at this stage. Often actors in performance send actions (verbal or non-verbal) out into the space and don't really see the way they land on their partner or take in the reaction. They simply do their next action. It is a bit like this exercise up to now. Two people sharing space, but largely doing their own thing.

Thick Space – Part 3: Playing Physical Actions on Your Partner

1. *As you continue to work with the space as your partner, expand your attention to imagine how you might send actions through the thick space to reach the other person in the room.* When you push against the space with your chest does it ripple across the room and land on them? Could it? How strong is the impulse by the time it reaches them? Does it fade over distance? How could you make it stronger by using more of yourself?
2. *In the same way, are the actions of your partner causing the space to move in your direction?* Can you imagine the way they change the shared space and how that might land on you?
3. *Begin to play with that. When the other person's action creates a change or ripple in the space that reaches you, allow your body to be moved by the space.* Where does it land on you? How strong is the impact? What ways does it change the shape and position of your body?
4. *As you play actions in response, do they land where you think they will? Is the impact as strong as you intended?*
5. *No need to get too close or to stay across the room. Play with different distances from one another. Take turns.*
6. *Let in what your partner is doing. Let it alter the exercise. You quickly discover that the space is no longer your partner – it is the material you use to play actions. It is the way you can impact/ change your partner.*
7. *Keep exploring different actions back and forth.* Steal ideas from them. If an action doesn't land the way you intend, try again using your body differently.
8. *Be careful that it doesn't become all hands and arms. Just as you did when first exploring with thick space, use all of yourself in each physical action.* Discover ways to impact different parts of them.

9 *As you work, be sure that your attention stays on them. Everything you do with your body should be in the service of trying to play an effective action on your partner. It is never movement for movement's sake. Your body is an action delivery device. The space is the medium of delivery. Your partner is always your target.* Be inside the curiosity, attention, action loop as you work.
10 *How do you know if it works? Have a desired way they might respond in mind. Be sure to stay with the action until you see some kind of response, then leave space for them to reply with an action.*
11 *Don't just perform actions for them to watch. Play the actions on them. Impact their body with the space.*
12 *As you work, are you planning your next action in advance? What if you allowed their action to inspire your next action?* Wait to see how their action impacts your body before discovering your response. Don't jump ahead in the loop.
13 *Work like this for several minutes.* Find a flow of actions between you. Don't worry about story or character or anything other than that action, curiosity, attention loop, and this physical exploration of it.

This experimentation with imagined thick space is an important step in physical actioning. It serves as a base for several other exercises with partners and groups and it demands continuous physical engagement and attention. Remember, your action leads to curiosity about their response, which turns to attention on them. You see and experience their reaction, which then sparks your next action. Thick space is an embodied exploration of that loop. It is slowed down so you can make more conscious observations moment-to-moment about your partner and yourself. With practice, you develop a physical conversation where your entire body is attentive and responsive to your partner and engaged in trying to affect your partner.

There are a few things to look out for as you work. First, guard against becoming a habitual receiver or doer. Is it really a back and forth? Are you a puppet, always receiving their input? Or are you always overpowering their action and not leaving space for response? It is important to note that this exercise doesn't involve actual physical contact. Resist trying to succeed in your actions by actually grabbing, pushing, or overpowering partners. The benefit of no contact is that everyone, regardless of strength or size, is capable of playing any action with anyone else. It levels the playing field and lets your partner trust you to play actions they might not if they thought an unexpected or unwanted touch might occur.

Avoid dead spots in the loop. At first, you may find that you play an action, your partner receives it, and then a kind of limbo occurs before they (or you) respond. It is vital to shrink that delay with practice. Often this is the point where you manipulate the impulse and habits rear their heads. Rather than truly responding to your partner you get clever or invent or suppress an action. You avoid the actual response you have to them. It takes practice to eliminate this delay, but you must. It happens all the time in performance. An actor hears their partner, acts between the lines, and finally delivers a manipulated and manufactured reply. It might feel great, but it isn't truly responsive. The work is not alive.

Revealing Your Palettes

Now that you and your partner understand the basics of working in thick space, explore playing physical actions with greater specificity. During the last few experiments you probably noticed some actions your partner repeated several times or maybe you recognized their Primary Action Palette start to emerge.

Learning Your Partner's Palette

1. *Look back at your personal Primary Action Palette from Part 2. Memorize those actions well enough so that they are easy to recall, even when your brain is quite busy with other things.* Remember, it is okay if your palette has changed since you first created it. You know much more now about physical actions than you did then and may realize there are better verbs to describe what you do.
2. *Take a few moments with your partner to find your way back into the thick space and playing actions with one another.*
3. *Once you have a good back and forth, begin to consciously add in some of the actions from your Primary Action Palette.*
4. *How might you play these within this exercise?* Remember when you expanded and contracted the size of them on your own. Perhaps that offers a clue about how you could fully embody them in thick space.
5. *Now that you have a partner, see if the action is landing on them and getting the response you expect.* Do they understand your intended action? How could you shift the way you use your body to make it clearer? What are you looking for in their body as a response? Are there other places/ways it might be showing up?
6. *As they respond with their own actions, are you letting them impact you?*
7. *Take it personally within the exercise.* By that I mean that you must play within the exercise like the back and forth is high stakes. Make it really important to you that you succeed at using the space and action to change them. If you fail, it should matter and make you want to come back stronger and clearer or shift to

another action sparked by that failure. Similarly, take the actions they are playing on you personally. If they seem to be playing "shakes," what do you think of that? What is the nature of the shake? Is it violent? Playful? Either way if you "take it personally" it demands a response from you through physical action.

8 *Rather than simply going down the list from your Primary Action Palette, play actions that seem to be an appropriate response to the action(s) coming from your partner.* For instance, if they play "adores," maybe "warms" from your palette seems right, or maybe "dismisses." There might be others that don't seem to fit.
9 *After you've played like this for a while, add verbal expression.* Try saying the action verb as you play it. Maybe say it over and over. Or make a non-word sound that feels right. Or say words that come to mind that fit the action for you.
10 *When you add speech, be careful that verbal and non-verbal communication have at least equal weight in the experiment.* Often as soon as speech comes in the rest of the physical engagement starts to diminish. Let the vocalization *strengthen* the non-verbal communication and support it rather than take over.
11 *As you play, find moments to interrupt your partner.* Maybe play a cascade of physical actions in a row to accomplish something. It is okay that it isn't a perfectly even back and forth like a tennis match. As long as you keep attention on your partner, stay available to receive their actions, and react to them, there can be messiness to the physical conversation.

What actions do they play as part of their Primary Action Palette? Can you name some of them? What does it reveal about how they see themselves? How do they receive you? What parts of your palette seem unclear to them? Do they respond to your actions differently than you intended? Good communication with collaborators is vital. Given how much conflict can exist inside a scene or between two characters, the more clearly you see and understand the other human beings in the room the better.

Sometimes when conflict or misunderstanding comes up in a rehearsal room, I ask actors what action they were trying to play in that moment. More often than not, the other person misread the action as something else, took it personally, and played an action that seemed justified in response. Then a vocabulary of those actions develops between people during the collaborative process. Even if you don't verbalize it, you can work to maintain curiosity in moments of conflict and ask yourself what action you received from them. Are you certain that is what they intended to play? What need are they trying to get met through that action? I don't think it is important to be friends with everyone you work with, but I do think you must to be able to differentiate between them and their role. Taking a bit of time to understand their personal action palette helps start the collaboration with clarity.

Bumping Relationship Palettes

Clearly shift your attention now to the situational self you created for this role in Part 3. As a reminder, just as you play different roles in different aspects of your own life, for the sake of the work on a role you create a temporary situational self. This likely includes and excludes parts of your own action palette, incorporates many actions based on the circumstances of the piece, and is the result of all that imaginative engagement with the text. Using the links between imagination and memory you generated a series of embodied events from the life of this person that formed action palettes and built this situational self.

Take a look at the Action Palette Diagram you created for this role as a result of that process. Remind yourself of how the various parts intersect with one another. Before you begin this next experiment, pay special attention to which parts of the diagram your partner's character would experience. Are there actions they would never see? Are there actions that you would hide from them? Once you each spend some time reminding yourselves of the ways these action palettes intersect, you're ready to begin the next experiment. We'll start with the expanded versions of these actions. Your partner will have plenty of time in rehearsal to see the way they are shaped by circumstances and by the Rules of the World. Here, they get an exciting glimpse of the strong unsocialized impulse underneath.

Interacting Relationship Palettes

1. *Take a few moments with your partner to find your way back into the thick space and playing actions together.*
2. *Once you have a good back and forth, begin to add in some of the actions from your Action Palette Diagram that you play with this other character.*
3. *Since you already explored these actions alone, how does having a partner alter them?* Remember to use the thick space to help expand them as fully as possible and provide more time to see their response.
4. *How are their responses (the actions they play back and how their body responds to you) different than when you were each using your own personal action palettes?*
5. *Do they understand your intended action? How could you shift the way you use your body to make it clearer?* What are you

looking for in their body as a response? Are there other places/ways it might be showing up?

6 *As you work, focus on this relationship and the circumstances. Remember to take it personally within the experiment. Play with high stakes.* Make it important to you that you succeed at using the space to get the action to land on them and change them. If you fail, it should matter and make you want to come back stronger and/or clearer or move to another action sparked by that failure. Similarly, take the actions they are playing on you personally.
7 *Rather than simply going down your list of actions, discover how these two characters inspire actions in one another. It is okay to discover new actions for this relationship.* Let actions that seem to be an appropriate response to what you receive from your partner come up. Practice having impulses the way this person does, rather than as you do habitually in life. Make sure you aren't simply sliding back into your own personal action palette.
8 *Let the space itself feel different. Create a shared thick space that makes sense for this relationship.* How thick or not can continue to shift and you discover the way your palettes intersect and what supports each action.
9 *After you've played like this for a while, add verbalization.* Try saying the action as you play it. Maybe say it over and over or make a non-word sound. Say words that come to mind that fit the action for you. You can add lines of dialogue that feel right for the action. Don't worry about the order of the text. Let the association between text, speech, and non-verbal communication go freely wherever you want. Even though the movement is slower, the speech can be normal speed.
10 *As you add speech, be careful that you don't lose the non-verbal engagement.*
11 *Discover how these two communicate.* Does one person always initiate? Does one of you always play long extended actions and the other short ones? Does one of you play primary actions and the other is always working in metaphorical actions?
12 *How is the rhythm of the physical conversation different with these palettes?*
13 *Does the role you're playing share actions from your own palette? If so, how do they play the same action differently than you?*
14 *Imagine this experiment as these two characters living in a world without rules, where their underlying impulses are free to be embodied through verbal and non-verbal means as completely, expansively, and openly as possible.*

Return to this experiment throughout the process of working on a scene or project. Your palettes of actions may change many times as you learn more about the role, your partner's choices, the aesthetics of a production, and the text itself. This work with collaborators informs and alters your initial choices from Part 3. Willingness to shift and discover within a process helps reveal habits about how you read a text; the actions you fear, avoid, or miss; and ensures that you keep curiosity and attention alive by always looking for what is new or changed. Often the more deeply you understand a role the clearer the palettes become and the more fruitful it is to return to this freely embodied exploration.

Working in this way early also reveals valuable information about how characters communicate and has the potential to provide ideas for verbal and non-verbal traits in performance. Try the two variations below to expand this understanding of how your palettes interact and how the relationship changes over time. They are an embodied way to explore moments that may not be present in the text but are vital for understanding how these two people communicate and interact through action.

Interacting Relationship Palettes

Variation 1: From Alone to Shared Space

1. Decide who is Partner 1 and who is Partner 2. Start with Partner 1 alone in the thick space.
2. *Partner 1: Play several actions from parts of the diagram that you would not play in the presence of Partner 2's character.*
3. *After about 2 minutes, Partner 2 enters the space without warning and immediately plays a physical action toward Partner 1.*
4. *How does their arrival change Partner 1's palette? What actions must be added and what must now be excluded based on your diagram?*
5. *How does the entrance impact the action Partner 1 was playing at that moment? Does Partner 2 notice the shift in Partner 1 when they enter?*
6. Do *different actions arise because of how Partner 2 entered?*
7. *Reverse this exercise and try starting with Partner 2 alone.* Don't assume it produces the same results.
8. *Did one person seem more like the owner of the space and one more like the interrupter?* How does each partner start the physical conversation upon entering? How does the other respond to surprises and unexpected arrivals?
9. *You can also explore this with less expanded versions of these actions.* How might you play those same actions through more everyday behaviors, activities, or language?

Variation 2: Finding Actions from Before and/or After

This variation works best with two characters that have a long arc of relationship over the course of a piece.

1 If your characters meet or develop a relationship during the piece, find a few actions that you played *before* meeting this other character but never play *with* them.
2 If your characters part ways/break up during this piece, find a few actions that you begin to play *after* that split but did not *during* the time you were with them.
3 If they both meet and break up/part ways during the piece then you can select a few for each version.

Start together in the room. Let the shape of the exercise follow whichever path the relationship does in the piece.

If you meet during the piece:

1. *Begin by playing physical actions alone in the shared space. Only play actions you would not play if they were present.*
2. *Find a moment to discover the other person and begin playing the Relationship Palette.* Who notices who? Who plays the first action? Do you discover actions that fit the first moments of this relationship that don't continue later? What are the first things you notice about the other person? What actions do they inspire? How do you get to know one another? Do you hide actions/information from them from before you met?

If you and the other person part ways during the piece:

1. *Begin physical actioning together using your Relationship Palette.* Again, it is okay to discover actions that fit the relationship beyond the ones you originally wrote.
2. *Discover the moment you part and play actions alone in the space. Who initiates the parting? Is it choice or an outside force? Pay special attention to those first moments after the split.* What action do you want to play on the space? Don't assume an emotional response. Focus on what you *do* in those moments after. Do you and your partner both agree on the moment the split happens or does one not realize until later?

For both versions:

1 *Be careful to avoid making this a dance about your relationship. You should always be using your body to do something to them. Play actions on them through the space.*

2 *Before and/or after your partner enters/exits play actions on the space as if the space was your partner.* Perhaps imagine another character you interact with in contrast to this partner.
3 The more your embodiment has a target and the more attention you pay to the success or failure of the action, the more it sparks your next action.
4 *Let go of an obligation to invent or be clever.* It is simpler than that and, therefore, harder to do.
5 *Just like the first variation, you can experiment with less expanded physical forms after trying the thick-space versions.* What do these meetings and partings look like through the text or as you imagine them? How do these same actions come through in embodied ways despite the less expanded form?

Impulses Submerged Under Text

The next several experiments build off of this core idea of putting attention on your partner and playing physical actions on them. So far you worked with action palettes and the relationship between you and your partner in a general way. While the moment-to-moment work in the experiment was specific, the next step requires looking at some smaller portions of text to ensure that level of attention and embodied responsiveness exists even when you know a great deal about what is coming next.

Writing and spoken language are processed differently in our brains.[3] You've likely had the experience of doing table work or private homework on a script and then feeling like you were completely starting over when you moved from the table to your feet. All of the exercises in Part 3 and these early exercises with a partner encourage you to use spoken text and non-verbal embodiment from the earliest moments of exploration. By the time you reach this part in the process, you already have a meaningful embodied relationship to these written words and key events from the text and this person's life. This continues in the following experiments as you and your partner work with the text on scenes and moments that you share.

When actors begin to work on their feet with text, speech often becomes so primary that they lose the strength of desire behind their actions and all of the ways we embody impulses beyond speech. Below, you'll use the idea of the Dials of Engagement from Chapter 11 to investigate the balance between speech and non-verbal expression, discover what actions best connect those two, and work to build a strong underlying impulse to action in performance. This relies on exploring that distance between the strength of your internal sensations and what is visible from the outside. You'll work to recreate the way strong desires get suppressed or redirected moment-to-moment by the

Rules of the World and the relationship between characters. It will help you understand how to take these expanded impulses and form them into something more recognizable as everyday physicality.

The Dials of Engagement – Partners

1. *Select an entire scene or about 2 minutes of dialogue from a longer scene. It helps if there is a fairly even back and forth.* That isn't to say it has to be a constant trade of single sentence lines, just that overall it works out to be even.
2. *Stand facing one another in the room and speak the text back and forth.* No furniture for now. Make certain you both feel solid about the words, so your focus isn't on recalling lines. There is no need to select an action for each line in the scene. For this exercise, let the actions be a point of discovery.
3. *Repeat the text a few times and notice both your own impulsive physicalization as you say the text and that of your partner. Do you gesture? Shift your weight? Step toward them? Guard against becoming rote or recited. Play the moments fresh with each repetition. Focus on what is different in your partner each time.*
4. *As you continue to repeat, think about the number of your internal dial.* Is it a 5? Lower? How strong is your need moving through you to say these words?
5. *Begin to turn up the internal dial.* It need not go all the way to 10. Find a strength that matches the circumstances of this moment. Make it important to get what you want.
6. *Notice the way this higher internal dial impacts the external expression.* How does it change the way you speak the text? How does it alter the rest of your body in gesture, movement, and behavior?
7. *Expand how your body engages to support your verbal expression and the internal dial. What else could your body do to make the actions clearer or stronger to your partner?* Can your gesture grow? Can it begin to include more of your body? Think of this as the strength of the internal dial *forcing* more to come out externally.
8. *Stay curious about how their body is changing. What physical actions do they seem to be playing on you?*
9. *Allow your body to respond just as it did in thick space, though you need not move slowly here.* Where does it land on you? How would the space move your body? How strong is the action by the time it reaches you?
10. *Be careful not to simply begin yelling your lines or just gesturing wildly with your hands. As you've practiced, allow more and more of your body to become involved. Let it become*

"unrealistic." Let your partner see the strength of the impulses behind the words. Give them a chance to witness what your underlying impulse is through your body before it goes through the pasta maker and gets socialized.

11 *Once your entire body becomes involved in playing these actions, allow the verbalization to diminish.* Keep speaking the text, but let it become a whisper. Then perhaps only single words come out – the ones that most capture the action. You might even replace the text of the script with the verb for the action you're playing. This isn't just changing the volume of speaking, but also how much you're relying on it to convey your actions.

12 *Experiment so that the physical conversation becomes more and more non-verbal. Eventually, it should be an entirely non-verbal scene.* If you feel lost about where you are, say a few words of a line to help your partner know where you think you are in the text.

13 *It is important this doesn't simply become a 2-minute dance or loop. Each time through (with or without text) remain open to changing the actions, clarifying them, or discovering an entirely new take on a moment.* Let your partner surprise you. They are different each time through. Look for the difference and respond to it rather than confirming your predictions.

14 *Once you've played like this for a few minutes, begin to reverse the process. Add back in some bits of text or words and then eventually entire lines.* Resist immediately returning to speaking the entire text.

15 *At the same time, allow the non-verbal communication of actions to slowly become more "realistic."*

16 *Can you keep the same strength of need and stakes even as you make this shift? Can you keep the sense of your internal dial turned up?* As you socialize the physicalization can you still feel the impulse start from somewhere inside your body besides your hands or arms? What internal sensations can you keep just as strong? How do you sense that in your body?

17 *How does the spoken text fill the gap left behind as your other physical expression diminishes?*

18 *Don't assume that you'll settle back into the same exact balance. Experiment with an equilibrium that allows as much non-verbal expression as possible to support the verbal expression.*

19 *As you watch your partner receive your actions, can you still see the way each impacts them physically?* Where is it still visible? Where do you experience their actions in your body? Can you see the physical impulses in them that are now being forced to come out verbally? Where in their body is that clearest?

20 *Take a final pass at the text when you think you've both found a balance of verbal and non-verbal expression where you are most effective at landing actions on your partner while allowing for the non-verbal elements to feel appropriate for the Rules of the World.* Sometimes that is more physically engaged than you realized was possible.

Life is full of strong impulses that are socialized, redirected, squeezed, suppressed, or otherwise formed, to successfully get what we need. Sometimes "Can you pass the ketchup?" really means "When are you going to apologize for what you said last week?" You know that many factors do the shaping, including our expectations of a partner's response. It is thrilling to get a peek at what your partner keeps hidden by turning up that internal dial and making it visible. You could do this exercise with entire scenes or even the whole arc of your relationship with the other character. Perhaps focus on key moments of connection or disagreement, when seeing what is under the verbal expression is most exciting or contradictory. As you continue to work on the script together look for how your actions land on their body. Recall the strength of your own fully embodied impulse when it starts to weaken, or use a more visibly embodied form if the scene allows.

Think about stage combat. It is a great example of physical actions between people. A punch or a stabbing is the embodiment of some powerful action that simply could not be contained solely in speech. The same might be true in other moments of non-verbal expression you either don't consider or where you fail to recognize the impulse. Your work is always physical. Those discoveries are more likely to happen if you prime that pump by taking the text and exploring the embodied impulses behind them.

Actions Through Activities

As you continue to rehearse, inevitably certain behaviors or activities become established through experimentation and repetition. Maybe a teacher or director gives you blocking or there is some task you need to complete in real time (i.e. getting dressed, or cleaning, or eating a meal). The risk is that these behaviors become divorced from your actions. Often the technical nature of these demands causes the action to shrink – manifesting less potently than it might. Instead, use the activity itself to bounce the action to your partner. The next experiment helps infuse the physical actions throughout your performance. Like the earlier experiment, Outside Behavior/Inside Action, you'll explore how required behavior for a scene can still maintain the strength of impulse to embody every action.

Turning Behavior & Activity into Action

1. *Select a section of text where one or both of you must do a specific physical activity or where you're struggling with how to link the behavior/movement you've been given with the actions that come from the text.*
2. *Begin doing the activity or blocking.* At first, it is better to use smaller segments of about 30–60 seconds.
3. *Speak any text from that section as you do the activity.*
4. *The first time through, simply play the moment as you have been rehearsing it, while turning your attention to all the ways different parts of your body engage with the prescribed activity.*
5. *Repeat this same section a number of times. With each repetition, think of a different action as you work.* Let the verb repeat in your mind. See how and when that action supports and strengthens the existing activity/blocking. Play the action *through* the activity.
6. *Remember, your partner is still the target of the action. The activity is just the tool you use to send the action to them.* Think of it like a bank shot in a game of pool. Bounce the action off of this activity to send it to your partner.
7. *After trying several, what action feels like the most consistent fit for this moment?* Are you selecting actions that support the circumstance? Does it help illuminate the activity rather than simply restate it? (Remember, the action "scrubs" doesn't illuminate the reason behind washing dishes. It's just a restatement.)
8. *Once it feels as strong as possible through this activity, begin to expand the physical action as you did in Expand and Contract. Do it over 5–6 repetitions.*
9. *Let go of the realistic version of the activity and let more and more of your body become visibly involved.* Let it grow into something that is unrecognizable as the activity, but very much the fully embodied version of the action.
10. *Continue to experiment with bouncing this action off objects, or walls, or whatever else is nearby, and onto your partner. You can also explore playing it directly. Feel the difference.*
11. *Did you drop speaking the text from the moment/scene? Could it help the strength of the rest of the embodiment? If there is no text, try saying the action aloud.*
12. *Once you've gotten it as expanded as possible, contract back down to the activity over 30 seconds.*
13. *Can you maintain the strength of the internal impulse? Keep the internal dial higher even as you channel it into a more realistic behavior?*
14. *Where in your body do you lose the sense of it? Where does it most easily stay clear to you?*

15 *Eventually you are doing the activity again, but with the memory and experience of how strong the impulse driving that activity can be.* Funnel all that strength of the action into this more "realistic" activity.
16 *Try this with a few different actions that might make sense for the moment.* Which one connects best to the circumstances? Which one seems to impact your partner most directly?
17 *Try the moment in the context of the larger scene.* Do the moments before and after help support the action through this activity? How could you change the actions you play around this moment to better support it?

Come back to this with many moments from the piece or the same moment as you understand it more deeply during the rehearsal process. You don't need to do this for every moment where you have an activity to accomplish. It helps most when you're struggling to connect with your partner through action while maintaining the behavior a teacher or director wants to see.

Be careful not to abandon what you've been directed to do easily. Investigate how actions can support the blocking and activities and, only after that seems impossible, experiment with what other activities better fit your existing actions. Of course, if your director or collaborators want you to generate these ideas, the experiment also provides a way to discover behaviors and activities that fit the world and keep your thought impulses embodied and communicative. In those cases, just begin with an activity of your own invention rather than one given to you.

You can also reverse the exercise. Begin with the expanded version of a physical action in thick space and contract it down into a smaller and smaller visible embodiment. Discover the activity or behavior the action becomes. Does it turn into walking across the room? Does it manifest in an adjustment of your clothing or fiddling with an object? An action has the potential to fill everything from leaps across the space to scratching your nose. Some actions link with activities quite naturally as you work in rehearsal. In moments where the connection is less clear, focus on embodiment and let the behavior/activity become the byproduct of the physical action.

Music and Physical Actioning

The next series of experiments all involve music and physical actioning. During my time as a student at the Theatre School at DePaul University, there was a class called "Movement to Music," taught by Professors John Jenkins and James Ostholthoff, that shared the core idea of these next exercises. They worked with music to liberate impulses and asked students to use the music to improvise a kind of physical conversation. We are beginning to understand that our perception of music is also embodied and that how people move in response to music can help reveal how they are perceiving it.[4] The experiments below build on that idea and link it to physical actioning, producing a series of tools for rehearsal with partners. Taken together, they explore elements of scenes or relationships that exist as subtext, find actions that you might not otherwise select because of your own habits, and alter patterns of speech and other physicality as you work to include the music.

Music and Physical Actioning

1. *Start with a song without lyrics and, preferably, one that neither you nor your partner know well.* It helps if there are frequent shifts in rhythm. Really good instrumental jazz is an excellent choice, but many types of music serve the exercise. It is worth having a few songs on a playlist so you can simply keep working.
2. *Start the music and begin the Interacting Relationship Palettes experiment, working with the space to play actions on your partner, finding your way into that physical conversation.*
3. *Rather than working in thick space, let the quality of the space be determined by the music.* That may mean you can move much more quickly, but you should still work as though you are using the space to play actions and change your partner.
4. *As you listen to the music, allow yourself to hear not just the rhythm, but individual instruments (like characters in a story) interacting and in conversation.*
5. *Pick one of those instruments and allow the notes and rhythm to become the impulse for each action you play on your partner.* It is as if the music is your verbal communication and the rest of your body is all the non-verbal elements of the action. For example, perhaps the piano notes provide an impulse to shift your ribs and send the action "pokes" across the space to your partner who receives it as a punch in their belly.
6. *Continue to experiment with your character's Primary Action Palette and the palette for this relationship.* You don't have to only use those actions but can also use the music to find new ways and new parts of your body to engage with these actions.
7. *Does this song fit a certain moment or event in the piece?* Does it link up with an event from the relationship that isn't in the script but you know occurred? What actions from those events can the music help you discover?
8. *Be certain that listening to the music doesn't divert your attention from your partner.* With practice, the music becomes a tool you use, part of the space, that helps you play more and clearer actions that land on your partner.

9 *How is the music changing their actions? What new actions are they playing and how is the quality of actions changing?*
10 *Are they linking up with the same instruments you are, or do they hook into different parts of the music?*
11 *It is good to let this experiment go on for several minutes (maybe three songs) to find your way into it and allow for waves of difficulty and discovery.*

It is important to remember that you are not dancing, despite the presence of music. Everything your body does is entirely about impacting your partner with the action, just like thick space. Your form or ability to stay on rhythm or the way your body looks to an outside observer is meaningless. All that matters is changing your partner through action.

Many people come out of this experiment the first few times pretty sweaty and not really able to articulate what they did or didn't experience. That's fine. The sweaty part will probably keep happening, but any confusion will abate over time. It is good to take a few moments after each round to make notes for yourself in your journal for the role about the actions you played or particular interactions that felt relevant to moments from the script. The music breaks up the patterns and rhythms of your own habits to generate greater variety. Some actors find that discovering a piece of music for their role (or each scene) helps determine how they use verbal communication, the pace of speech and movement, and other internal experiences of the world.

The next several pages are variations on Musical and Physical Actioning and offer many ways to explore events. You and your partner will experience surprise and change from the music with each one. They force you to keep your curiosity and attention strong even though you know what comes next in the text. They all begin with the idea of two partners in a room using their bodies and space to play actions on one another. They all expect that you continue to use the music as a tool for changing your partner as you did in the experiment above.

Variation 1: Single Scene

Select a song that you and your partner think captures the arc of a single scene. Use the music to play actions moment-by-moment. Feel free to add single words, phrases, or entire lines as they enhance the attempt to play the actions. You may use most of the text or none at all. You might also replace text by speaking the name of the action you're attempting to play. Be sure to keep the physical form of the actions expanded rather than retreating into simply "acting" the scene. Let the music help.

Variation 2: Relationship Song

Select a song that reminds you of the best moment in this relationship. Let it be a song this person would know from the time period and location of the piece. Don't tell your partner what the song is in advance. Play it out loud twice: once while you work through a scene with physical actions alone and once while you speak the text and allow the non-verbal communication to be more realistic. Allow the song to influence your actions as you work together. Does the scene you selected contrast with what the song offers? Does it support this song memory? How close is this scene to the best moments of this relationship?

Variation 3: Before or After

If the relationship between your two roles has a long arc, select a song that fits the beginning of the relationship and one that fits the end. Play them one after the other and explore through physical actioning how the palette shifts in the jump from the start to the finish. What actions disappear? Do you play the same actions differently after there is much more history? Be careful not to allow the music to become mood or simply generate sentimentality. The instruments and rhythms are all driving your actions, not simply a score for your movement work. Don't be afraid to verbalize the text or name the actions you're playing out loud. Include vocalization and verbal forms of action playing whenever they might support the exploration.

Variation 4: Song and Entrance

One actor starts alone in the space and plays a song that reminds them of the other character. They stand facing the entrance to the room. The second partner enters after several seconds to discover the person and the song. (Don't tell them in advance what the song is.) As soon as the door opens, discover who has the first physical action and begin to play from there.

Variation 5: Volume

Play out loud a song that fits your sense of a scene or the relationship. As you work together with a character palette or a specific scene, Partner 1 has permission to turn the volume up or down based on how effective they think Partner 2 is at landing their actions. Be sure to reverse this. You can also have a director or other third person control the volume as you and your partner work together.

Variation 6: Competing Songs

The same as *Variation 5: Volume*, except both actors have a song they selected playing out loud at the same time. A director or other third person turns the songs up or down depending on who they think is most successfully getting what they want/need from the other person. The volumes could stay quite consistent or change constantly. Be careful to notice the difference between status and successful actions. One partner can play very strong actions, but the other partner might still be much more effective. Watch for what they seem to want, and how completely they achieve it.

Variation 7: Singing

Pick a song with lyrics that fits the relationship or a specific scene. As you work through physical actioning, sing to your partner, mixing the singing in with the text.

Variation 8: Anti-song

Is there an anti-song for this scene (just like you have an anti-palette)? Something that is completely out of context, period, setting, and rhythm? In what way do your actions and the scene feel different as you incorporate that music through physical actioning? Are there moments where the contrast opens up something new? Are there ways that it invites different actions or a fresh perspective on a moment?

Variation 9: Earphones

Each partner listens to a song privately (with earphones) that feels like it matches the relationship or a specific scene for them. Don't tell your partner what the song is. Now play together with physical actions as you listen to those songs. Can you see the way your partner's secret song changes their rhythm and the nature of the actions? Are there moments when you sync up? Where does it live in their body differently? If you use only one earphone and leave the other ear free, this can also be done while verbalizing the lines.

Variation 10: Switch Songs

Try physical actioning a scene while switching earphone songs with your partner (without telling them the song in advance). See the scene from their character's point of view.

Variation 11: Shared Secrets

Start with *Variation 9: Earphones*, but share one earphone with your partner during the scene in moments when you want them to hear your song. When would sharing this secret be most effective for achieving your action? What is it like to suddenly be let in on their secret? What happens when they take the earphone back? Do you want to keep their song or reject it?

Variation 12: Their Song

Like *Variation 9: Earphones*, but select a song for your role from your partner's perspective. What song would their character pick for you? Try listening to that while you play with the scene.

Variation 13: Third Wheel

Like *Variation 9: Earphones*, but listen to a song that reminds you of someone other than your partner (another character, previous lover, etc.) as you play actions with this partner. What does this secret do to the actions between you and your partner?

Variation 14: Personal Song

Like *Variation 9: Earphones*, but experiment with a song from your own life that has relevance to this scene or set of actions. Maybe it is from a time where you experienced a similar event. If not, perhaps the song sparks your imagination to fantasize or invent similar scenarios to the scene.

Those variations cover a lot of ground and give you many ways to experiment with the relationship, text, and circumstances while staying embodied and focused on your partner and actions. I'm certain that the variations offered here will spark your imagination and generate many other ways you might play with physical actioning, music, and text in combination. Each role you work on offers a new opportunity to use these variations based on the circumstances and the relationship. Perhaps the setting or time period of a project will open doors to new genres of music to explore. I encourage you to invent variations as projects allow.

Don't Forget the Box

One warning about this work is not to stray too far from what can be known about the role. As we discussed, every world has rules. Every production completes that box by further defining a set of rules for that version of the world. Sometimes this work helps you and your collaborators create those rules, but other times they are well defined before you join the project.

The point is not to turn every play or project into an abstract movement piece set to music. The point is to explore the thought impulses through verbal and non-verbal embodiment so that you can understand *how* the Rules of the World shape, constrict, form, and alter the actions you want to play. These experiments help you ensure that you are full of potent desire to play strong actions even when there are tight restrictions. In fact, the tighter the Rules of the World the greater the value in understanding what you are holding back. The experiments also help you to see your partner's strong impulse to action that is constrained by the rules of that shared world.

Those impulses, however, should still make sense for the piece. The actions in your palette and that you discover through these experiments are built on the foundation of work you did in Part 3 and in keeping with the story. If your character is a non-violent Buddhist monk and through these experiments you keep playing actions like "eviscerates" or "bludgeons" that may be right, but do check that it isn't just that you're enjoying the liberation of the exercise. It is a useful exploration when it expands out from the text and the world you create with collaborators.

Now that you've done these key experiments about attention and action in space with another person, you and your partner can also investigate some of the experiments from Part 3 together. What would it be like to walk key events from these lives together? What about exploring the shift in palettes from setting to setting and person to person with your partners there in the room? The same experiments that helped you develop early ideas about the role can now help you and your collaborators refine those and build embodied memories as you continue to rehearse.

So far in these shared experiments you have focused on two-person work. Some of the work easily adapts to more. While many scenes have two people, new complications arise when the room has three or more. All of them play actions. All of them have needs they want met. All of them have individual histories that impact the actions you play together. It's an exciting challenge and one we navigate all the time in life as we try to accommodate and act upon a variety of people seemingly all at once. Let's look at the opportunities that presents in your work.

Notes

1 Spolin, Viola. *Improvisation for the Theatre*. Evanston, IL: Northwestern University Press, 1983, p. 81.
2 Chekhov, Michael. *On the Technique of Acting*. New York: HarperCollins, 1991, p. 45.
3 Rapp, Brenda, Simon Fischer-Baum, and Michele Miozzo. "Modality and Morphology: What We Write May Not Be What We Say." *Psychological Science* 26, no. 6 (2015): 892–902. DOI: 10.1177/0956797615573520.
4 Leman, Marc, and Pieter-Jan Maes. "Music Perception and Embodied Music Cognition." In *The Routledge Handbook of Embodied Cognition*, edited by Lawrence Shapiro. New York: Routledge, 2014, pp. 86–87.

13 With Three or More

Three's a Crowd

There is something inherently theatrical about a group. Audiences lean forward when two people are in a space together and a third person enters. Questions immediately arise. In this chapter we'll look at how to explore physical actioning with groups of three or more. It provides exciting opportunities to expand your sample size for how actions impact people differently. Once again, if you're not working in a class or rehearsal, I encourage you to have one person serve as teacher/director for each experiment and read the instructions. It also helps to have someone who can offer insights watching from the outside.

When I teach, I lead many of the individual and partner experiments from the previous chapters with the entire class present. Seeing the way others work together or hearing the actions classmates pick for moments vastly expands your own palette. I tell students all the time to steal exciting ideas or ways to physicalize actions and try them out for themselves. The collective space of an ensemble or classroom teaches you a lot even when you are not the one up working. That said, participating in group experiments is a unique and exciting part of this work. If you're reading this book on your own, I hope you get the opportunity to explore the ideas in this chapter with a group and see how different it is from working one-on-one. We began the last chapter exploring the dynamics of the partner relationship. We'll begin this chapter examining the interaction of the specific group in the room at this particular moment.

Learning the Group

Start with a number of members of the group walking around the space together. Remember, this work is best done with movement mats covering the floor. If there are so many people that it is hard to move, work in smaller groups. Keep them random to avoid self-selecting friend groups. There should be enough space so everyone can move in large and expanded ways without worrying too much about accidentally hitting someone.

1. *As a group, allow yourselves to discover the thick space. Rather than just inventing your own, see if you can discover it together.* How quickly does it start to resist movement? How does everyone begin to match one another? Are certain people not in the same space as the rest of the group? Focus on you matching the group's thick space, rather than correcting others when you think they aren't in the same thick space as everyone else. What is the shared thick space you create together?
2. *Do more than just walk in thick space. Experiment playing expanded physical actions with the space as your partner.* Work alone for now, simply getting warmed up and ready to turn your attention to others.
3. *Make eye contact with someone else in the group. They will be your first partner.* How do you find your first partner out of this larger group? Were they right next to you? Across the room? Who did you *not* select? Why?
4. *Now begin to explore the Learning Your Partner's Palette experiment from* Chapter 12. As a reminder, you explore your personal Primary Action Palette and work to discover how it interacts with this specific person. What other actions come up that are specific to your relationship with them?
5. *As you engage with them, who plays the first action? Who starts this conversation?* Spend a little time with this person. Let a physical actioning dialogue form.
6. *After a few minutes, have one person play an action that ends this conversation.* What kinds of actions accomplish that? What feels final? Who is the one that plays the last action?
7. *Staying in the thick space, find a new partner.* This can be challenging. Sometimes everyone else is occupied. How do you join a conversation? Do you interrupt? What actions do you use to draw someone away from their current partner?
8. *Continue working like this. Some conversations might be quite brief while others go on for a long time.* Over the course of the exercise try to work with everyone at least once.
9. *Notice the way different collaborators respond.* Can you do this without judgement? Try playing the same action with multiple people. How does it land on their body differently? What changes do you make in playing this action with this specific person? What differences are there in the action you get back? No two people will respond the same way.
10. *Don't always plan your actions in advance. Be available for what comes your way.* Don't forget, it is fine if you can't always name the impulse with an action verb. Let the response be physical and maybe you'll discover the name for it. Sometimes you won't.

This early exploration of embodiment with a group establishes a courageous jumping in and playfulness for the work that comes next. Certainly, the actions and the way you play them depend on whether or not the actors already know one another and are comfortable working together. If not, it helps to start with some of the solitary exercises. I don't recommend having people work in pairs while the group watches at first. There is a liberating anonymity in being a part of a large group only watched by a teacher or director. You can experiment and flail around and figure it out without the pressure of performing. Everyone is in the same leaky boat.

Move quickly into exploring this same exercise with the Relationship Palettes below. Your memory of your collaborators and their choices are fresh in your mind and the differences between their personal palette and the ones for their role become clear and exciting to witness.

Interacting Relationship Palettes – Group

Before you work, review your Action Palette Diagram for this role. Look at the ways that actions become available with certain partners. See what actions you exclude with others. It is okay to make new discoveries as you experiment, but start with the diagram as your guide for the work.

1 *Start by working like you did in Learning the Group above, but this time only work with your Primary Action Palette for this role.*
2 *Work for a time with each person and then change partners. Discover the ways each new collaborator responds to these actions.* What does it change/liberate/restrict in your own impulses to respond?
3 *Once you've played with that primary palette for a while, begin to change partners every few minutes.*
4 *As you work, think about the history and relationship with each new partner's character. Allow yourself to shift the actions person-to-person so they match your relationship palette for this partner.* If you don't have a relationship palette for this partner, stick with the Primary Action Palette for the role.
5 *Notice what happens when one physical conversation ends and another begins.* What actions do you revert to? How do you adjust palettes for this new partner? Do new parts of your body become available to initiate? Does your sense of the thick space change?
6 *Keep working like this, trying to cross paths with everyone at least once.* Remember, when you don't have a Relationship Palette for someone, revert to your Primary Action Palette. As you work, discover new actions you might reserve for that person or eliminate certain actions that don't work in your relationship with them.

Once you explore the variety of actions in your diagram and how they interact in practice, try the following variations. They focus on the experience of one particular actor/role in the group and how their Action Palette Diagram works.

Variation 1

Set up the space much like the Action Palette Diagram one of the actors created. They stand in the center while other actors stand around them marking where they are in the relationship or setting palettes. The actor in the center starts physicalizing their Primary Action Palette and then begins engaging different partners standing on the edges. Notice if a pattern develops of who comes after who? Who do they go to first? Which action initiates the dialogue? How does the thick space change, depending on the partner? The actors on the outside keep exploring the actions in their Relationship Palette for the person in the center, even if they don't have their attention at the moment. Feel free to speak text or say the action verb you're playing. If it becomes too focused on verbal actions, restrict it to single words or silence.

Variation 2

Like Variation 1, only the actors on the outside are free to move and attempt to use their Relationship Palette to draw in or reject the person at the center. They can also interrupt another conversation if it makes sense for their character and relationship. Similar rules apply about speaking.

Music and Groups in Action

A group version of the music and physical actioning experiment is a valuable way to produce greater variety. An initial version for the group can simply be an expanded version of the experiment offered for pairs. That said, when working on a specific scene or project the introduction of music that illuminates the larger world of the script is immensely helpful. It's worth spending some time in advance finding songs that are from the period, location, and culture(s) represented in the script. Some music that addresses the world of the characters broadly is good, but it also helps to find songs that particularly speak to certain key relationships or individual roles.

Sometimes I ask actors to bring in a few songs that speak to them about the role. These can be from the time and place of the script, but can also be music they already know and like that connect them to the story or circumstance. Contemporary music, especially songs that the company knows well, can be a trap. Often the experiment merely becomes a dance party or a sing-along. Those are fun, but they happen not to be the point here. Still, many actors come up with surprisingly evocative contemporary music that supports exploration of the piece and still serves the physical actioning.

Music and Physical Actioning – Group

1. *Mirror the process from Music and Physical Actioning from* Chapter 12. *Begin by using a song or two to get the company warmed up into the experiment.*
2. *Shift to songs that fit the world of the script.* Notice how this music opens options for new actions or ways to express the selected actions.
3. *Do certain songs make more sense to explore with certain partners?* Find that person and make it happen. Hopefully they agree!
4. *If a particular song seems to match a moment of the text, try speaking a bit of it aloud as you work.* How might the music strengthen the quality of the actions?
5. *It is okay to work in groups of more than two at a time.* Some of the music might encourage or remind you of group events from the text.
6. *Explore how you play a single action on two or more partners.* Do you just keep switching your attention? Could you play a single action on multiple people at once? How does the same action affect different people in the group differently?
7. *When you're done, make some notes about possible new actions you discovered through the music.* Did the music help you find actions that might exist inside the rules of this world that you didn't think of on your own?

Contracting to Realism

While many actors benefit from expanding out the embodied impulse for physical actions in this way, you may struggle to link how these versions of the actions would manifest in the performance itself. The expansion strengthens your visceral sense of the action and makes the way it impacts your partner clear, but if you are not going to be moving around in thick space to music during the performance, how will these help? The answer partly lies in the Expand and Contract, Outside Behavior/Inside Action, and Dials of Engagement experiments you did previously. The experiments with the expansive versions of the actions build a memory of your embodied engagement. When done in pairs or groups, your partner gets a chance to see the strength of that underlying impulse, whatever version of the action ultimately ends up being visible in performance. This allows you as partners (or as a company) to continue to adjust the scale of embodiment for the production as well as moment-to-moment in the work. Still, it is worth actively experimenting with exactly where that sweet spot is for the visible embodiment for your particular scene or project. These next exercises actively explore the level of visible embodiment for actions in a group.

Let's start by talking about distance. If you can, create an area with mats that is the same size and dimensions as your performance space. Often there isn't enough space in a rehearsal room to mirror the stage or set, but get as close as you can. It helps if you can be in the space itself, as it allows the director or other viewers to be at the distance the audience might during performance.

Building the Rules

Select a number of sections of text for this experiment. It's best to make sure a variety of actors/roles are represented and that there are different sized groups.

1 *Start by simply speaking the text and walking/traveling around the space in whatever way feels realistic or truthful.* Unless it is a very long scene, you'll likely repeat it a few times. When you get to the end, simply start again immediately.
2 *As you play the scene you are playing actions moment-to-moment, even if you haven't selected them for every line.* Those actions come through via speech, behavior, and activity. It is okay if you can't name them all.
3 *As you work, the director/teacher/observer periodically calls out a number between 1 and 10 for the level of expansion the actors should try.* While sometimes it helps to jump from numbers that are far apart, often simply adjusting by one or two numbers produces a strong shift and one that actors can understand within the scene.
4 *If the director/teacher repeats a number they just said, that indicates that you are not yet fully arrived at that number.* It is often a sign you have not expanded or contracted enough in your embodiment of the physical action. This also helps you create a shared sense of what the numbers mean.
5 *Don't be afraid to go past what you expect is "realistic" for the sake of the experiment.* Go just up to and past the edge of it in either direction.
6 *Once you've done this for several minutes, the director/teacher begins to also call out "Distance" followed by a number 1 to 10.* Use the same sense of scale here. 1 is very close to your partner. 10 is as far away from them as the room allows.
7 *Continue to experiment with how these two scales work in harmony and contrast.* What is a section of the scene like very close together and physically expanded to an 8 or 9? What about very far apart but shrunk down to a 2?
8 *Experiment with each actor having a different level of embodiment, too.* What if one person is an 8 and the other is a 4? Does it reveal something about relationship? The actions available to them?

Working in this way with several scenes and varying number of actors, you discover exciting images for production and build different ways in which each character can and does embody their actions. It is important that you don't attempt to land on a single answer for the entire production or for every character. You'd make a very boring project if everyone was at a 5 in their embodiment and a 6 on their distance all the time. It makes me sleepy just thinking about it. What you are really assessing is the range and scale for the rules of this particular world. You're building the box that holds what is possible and how to express the physical actions for this world. Part of the excitement of watching a production is seeing who pushes against the rules and how. Think about someone standing still in the middle of a dance floor or someone dancing in a porcelain shop. Both people are doing reasonable activities, both are mismatched for the setting. Without understanding the rules, you as an actor cannot know and respond when a scene partner crosses that invisible line.

This exercise may also uncover actions you thought were part of your palette but simply cannot be played within these parameters. On the other hand, you may also discover new actions that are especially potent within this world. To be clear, a highly physically restrictive world can still contain inherently expansive actions like "eviscerates" or "smothers." In fact, the restrictions might make them more possible. The question is really about what physical actions seem to have the clearest and most powerful ability to land on your partners as you and the rest of your collaborators establish the physical world of the project.

The Space is an Action Magnifying Glass

While your performance space might be empty, there is a strong likelihood the world includes some objects and furniture. It is important that you recognize these items as potential tools for playing actions on your partner. As you've seen, some actions literally use an object. Imagine playing "challenges" by handing someone a piece of paper with your phone number on it. The piece of paper becomes a part of the embodied version of the action. We use all sorts of objects as extensions of ourselves to interact with others. Some of these are weapons, but many are tools of various kinds. In fact, there is strong science indicating that these tools get temporarily incorporated into our sense of our own body, or "body schema," so that we can use them better.[1] They become a part of our embodied selves through which we play actions.

We also explored times that you use an object to play an action more indirectly. Imagine someone scratching at a stain on their coffee mug while they tell another person that they have to leave town for a while. Maybe they're playing "softens" because they anticipate their partner will be upset. There is no eye contact. They seem to be playing the action on the mug itself. In truth, they are bouncing the action off of the object to reach the other person. These moments are much like the example from Part 2 where you found moments from life when you did an activity while playing an action on someone else.

This same dynamic exists with furniture. "Blocking" is best motivated by how effective the move is for playing a physical action. Maybe you want to be further from the person in order to play an action with strength. Some things are better said sitting next to them. Simply sitting down on a chair can be the embodiment of an action. Rather than differentiating between blocking and action playing, think of them as parts of a whole. Your movement, behavior, and activities are a result of how you can best and most effectively play actions within this world.

Objects and Furniture

Set up the performance space (or a rehearsal space to as similar a size as possible) for a particular scene with the furniture and objects you have for rehearsal. It is helpful to have a few objects that might be in that room but are unused during this scene right now.

1. *Each actor selects an activity to do during the scene.* Let it be something you can actually do, that makes some sense in the world of the script, and be sure to have the materials to do it. It is good to pick a section of text no more than about 5 minutes, since you'll do it multiple times.
2. *Take turns playing the scene while one of you does your activity and the other does not.* Both are free to move locations and change positions within the room.
3. *How does having an activity (or not) alter your actions, their strength, and how you play them?* Are certain actions best played *through* these objects?
4. *Experiment with not looking at your partner or limiting yourself to 1 or 2 moments you're allowed to stop the activity and look at them directly.* Where do you use those precious limited glances? Careful not to always pick the strongest actions to do it on. Challenge yourself to use the activity and objects to your advantage rather than an obstacle.
5. *How does the actor without an activity deal with the indirect nature of the actions coming from their partner?* How does it change the actions they play in response?
6. *Try a version without a prescribed activity but where each actor must play an action through at least one object or piece of furniture during the scene.*

While there will be many moments where simply looking at your partner is key to effectively playing an action, you'll be surprised by how often the physical world around you offers tools to help. Even if you discover just a moment or two for any given scene, you expand substantially what impulses you can follow as work on the project develops and as you explore with other partners.

New People? New Space

It's difficult to track the impact of new people entering or exiting when you work on a group scene. As you know from developing your Action Palette Diagram in Part 3 and the Relationship Palette exercises earlier in this section, certain actions become possible to include or exclude when other characters are present. It's helpful to experiment a bit with scenes like this to ensure everyone is on the same page about how the world shifts in these moments and for whom.

In and Out – Group

Select a scene with several people. It tends to work best with five or more and that at least one person enters or exits during the scene.

1. *Begin with everyone who is in the room at the beginning of the scene.*
2. *Each actor looks at, points at, and says an action that they play with each of the other people in the space.* It doesn't matter if the other person is looking at them. This can be done all at once. It may make a kind of cacophony of words for a few moments.
3. *In the order that it happens in the scene, have each person enter or leave the space.*
4. *As they do, everyone else should say either an action that is now available or "No," followed by an action they now must exclude.* For example, if your boss enters the room and you only play "elevates" with them, you would say that. If they left the room, you would say "No elevates."
5. *The person entering or exiting says the physical action they think they play as they enter or exit.*
6. *If the scene includes revealing significant new information (like an engagement, a death, a threat, a secret, etc.) also have the person speak that line.*
7. *Everyone else should then say how that information changes their available actions.* For instance, if someone comes in and reveals they are getting married and another character in the room is in love with them, maybe they would need to say "No seduces," or maybe "Dismisses." On the other hand, perhaps they just keep

trying anyway, in which case that might not be the result at all. If nothing changes, you can say "No change."

8 *Once you've done this, play the scene with the text, exploring those actions or their disappearance as you work within the full scene.*

This experiment also reveals to a director how the actors experience and receive the information. Maybe they're making it too important or perhaps they aren't realizing the impact it has on them. If so, directors can offer the actions that must shift. One variation on this is to play the scene and, when these moments arise, the physical actions must be played with fully expanded embodiment. Let them feel like a massive spike in the life of the group. The clearer that sense of change, the more that shift can be directed and shaped into behavior and activity along the way.

Actioning Line-by-Line

While you began your individual work on the script by reading and walking the role in Part 3, it has been a while since we discussed a rigorous look at the text itself. Much of the discussion about using actions tends to begin by looking at the script line-by-line. Several versions of applying actions to text exist. Some ask that you apply an action for every sentence. Others say every punctuation mark, including commas. Others are more flexible in the length you can play an action.

I'm not quite as strict as some in this area. I think, especially starting out, it is incredibly valuable to assign physical actions line-by-line for a number of key scenes with different partners. It is good practice, it builds you vocabulary of actions, and it demands specificity where you might lean on tone, emotion, or attitude. That said, the actions you select must not become a religious text you follow faithfully. They are a starting place from which you and your collaborators explore and discover. Selecting actions line-by-line helps you make the offers grounded in the text and the Rules of the World clear and specific. Your collaborators benefit from that, even if those offers turn out not to work down the road.

By this point in your process you have explored a wide variety of physical actions for this role with and without text and in relation to other actors and characters in the world. You have a massive number of embodied memories from these experiments. Rather than sitting down and turning this next step into table work, the experiment below asks you to apply the same skills of embodiment to this detailed examination of the language in the text. It folds several ideas from your previous work into the practice of assigning actions to each line. I put this work late in the book because it is best to wait and do the line-by-line work together with collaborators whenever possible. If you and your classmates or ensemble explored the text in all of these embodied ways, your capacity to think and name the action impulses collectively is exponentially bigger than if you tried this after a few reads of the text on your own. That said, you can also action line-by-line on your own for scenes and projects if needed. It is still a valuable way to apply this vocabulary with moment-to-moment detail when needed.

Select a two-person scene that is at least a few pages long and where the other character is one of the key relationships for your role. Take your notebook for this role and find where you wrote observations during your first read for this scene. What did you expect would happen as an audience member? What paths did the writer send you down? Are they purposely misdirecting the audience? Where were the discoveries or new information? What physical actions or ideas did you write down as you walked the role? Were there moments from the history of this person you explored that involve this other character? Are those events at play in this scene? Really examine the context of this relationship and what happened between these two people before this moment. Now look at both your Primary Action Palette and the relevant Relationship Palette. With this context and set of responses fresh in your mind, try the following exercise with this scene partner.

Actioning the Scene

1. *Each partner makes two columns on a blank sheet of paper. One for the line, the other for the actions.* Keep it nearby and easy to reach on a table or music stand.
2. *While standing, read the scene out loud all the way through once. Make notes as you go of actions you think you played.* No need to catch every moment. Just make notes when something strikes you.
3. *On the second read, work together reading a few lines, and then go back and try to name the actions being played line-by-line.* Don't be afraid to talk around it, try out multiple actions, or even let your partner offer their thoughts about what they are receiving. What do your impulsive gestures tell you about the action? Can your partner describe what they see you doing?
4. *Use the information from your partner's choices.* How can your choices offer them resistance? Are their physical actions revealing a different view of the relationship or scene? Give them something to resist.
5. *From time to time, or if you're struggling with a moment, expand the impulse and allow a more visible physical form. Try it in thick space. Repeat it. Find the form of the impulse and then name what you discover.*
6. *Continue doing this line-by-line together until you've completed at least 20 actions per person.* Even better if you can do the entire scene. Are some actions played without text? In a silence or pause?

7 *Notice when and if physical actions from your Primary Action Palette show up.* What about actions that came to mind when you first walked the role? Do those actions still fit?
8 *Be careful not to simply deliver physical actions that set your partner up for success or that are too aware of what comes next. Remember that part of the audience's excitement when watching is seeing you try and fail.* What does the text support? What is in keeping with the Rules of the World? The status? What fits the palette for the relationship?
9 *Don't simply reuse the same actions over and over.* If you're repeating an action more than once in a scene you might think about the Cascading Actions. What small shift can help explain why you keep speaking or repeating or attempting similar things?
10 *You've now generated an initial Score of Actions for the scene.* Way back in Chapter 3 you did a very simple version of this.
11 *Keeping the sheets nearby, explore at least the first 20 physical actions in thick space together. Speak the text and play the action in the most expanded form.* If you get to the end of a line and your partner doesn't respond, keep playing the action and repeat the line or words from the line that help strengthen it. It helps to have a third person like a fellow actor or director present to feed you lines or actions, so you don't have to keep referring to the page.
12 *Do this a few times, finding clearer flow action-to-action and line-to-line.* The space can become less thick as you find the flow while keeping the expanded embodiment.
13 *With repetition, you can also explore contracting the embodiment back down.* Careful not to do so in an entirely even way. Some actions and moments might invite a more expanded embodied expression of the action. Don't be afraid of that even as you work toward a level that fits the Rules of the World for this text and production.

Just as you found when first exploring in thick space, it can be easy to simply stay a receiver of your partner's actions for too long and miss the impulse to reply. This is usually the result of either a desire to manipulate and plan the action you're about to play or your partner's action being too weak or unclear to generate a response. Whatever the reason, even in thick space the flow of action, curiosity, attention, and re-action should be seamless. Certainly, some actions take longer to execute than others, but there should not be a gap or dip where the first partner has long since played out the action and it is starting to decay while the other partner continues to receive. In all likelihood, the actual impulse to respond long since passed.

After you do this with a two-person scene, try one with three or more people using the same basic structure for the experiment. You'll discover how the earlier work on entrances and exits along with the specifics of your Action Palette Diagram become a valuable tool for understanding and shaping this detailed investigation of text. Even your sense of the Rules of the World may substantially shift depending on who is in the scene at a given moment.

As you can see, working in this detailed way with a text is most effective when you and your collaborators all do the early embodied investigation of the role. You bring all of those experiences and ideas into the moment-to-moment work and are primed to offer exciting actions grounded in the circumstances and shared world supported by a deep understanding of the role and the text. Even if that isn't possible, don't hesitate to action the text (or key scenes/ moments) on your own.

In Performance

I always wince a little when I hear people talk about "forgetting your work" when you get into performance. I think I know what they mean. After all of the rehearsal and investigation and direction and interrogation of the piece, good performance needs to have a feeling of lifelike spontaneity and liberation that building a performance often lacks. Still, I worry that many actors misunderstand this as an instruction or invitation to abandon what they've learned in favor of whatever impulse comes along. Some of those are less about the project and more about anxieties and the safety of old habits and the inevitable desire to be liked by the audience.

I've never driven a train but my friend, Mike May, has. I asked him about the difference between learning how to be a locomotive engineer and driving a train day-to-day. He said:

> Running a train involves a lot of thought and quick decision making when you're running through busy territory, and in the wide-open cornfields can be quite boring. It all depends on your surroundings. [There are] really two sides of the job. One is actually running the locomotive, power, brakes, handling of the train, so on. That's the part that becomes second nature to skilled engineers. The other half, and more important half, is getting a train over the road safely. This is where knowing the thousands of operating rules and how they all apply comes in. Knowing how your train interacts with all the other trains and workers out there. That's the part that changes day to day and minute to minute ... having that strong foundation is what leads you forward almost on autopilot, it's all new things to think about while your "rehearsal" time is why an engineer can run the train without thinking and think instead about the train getting over the road.[2]

This seems like a really beautiful metaphor for the difference between all the work offered in this book and how you engage it in performance. Taking the time to explore a role in an embodied way from many different angles and with tremendous detail and rigor does not rob you of the ability to work on impulse in performance. Just because you are not thinking about *all* of that work moment-to-moment in performance also does not mean you "forget it."

In truth, you are working from conscious incompetence to unconscious competence over the course of rehearsal. You form associations and memories and make connections to your own experiences. This takes time and attention and can be a genuine pleasure. The first rehearsal or even the twentieth aren't the performance. They are part of the investigative journey. Then, during performance, you can return to the simplicity of the loop. You can play your action (whether you're focused on a specific verb or not), follow through with curiosity and attention, and see what comes back. This kind of "autopilot" isn't an abandonment of what you learned in the process. It is turning that skill and practice toward what is different and changing. It helps you focus on what is down the tracks.

At the same time, you're not *unaware* of driving the train, you're just incredibly competent at it. When you run into a moment or scene after several performances where something just isn't working, you have the tools to shift the physical actions, bring it back into a conscious state, and change the way you're playing it. No matter how much the same it feels, the repeated moment or gesture is, in many ways, occurring anew each time and directly connected to your body and the environment in that specific moment.[3] You can check in about the status or how expanded or contracted it is. Maybe you can find a new place from which to initiate the impulse. Or maybe you discovered something new with your partner and it requires a shift in actions. This is you playing with the brakes and power and handling of the train on this specific trip. Sure, it can go the same smooth way time after time, but when it doesn't, you can bring back the tools you learned and reapply them in a conscious way. There is also tremendous pleasure in the ability to make these shifts and adjustments on the fly.

Because our brains are such phenomenal predictive machines there are real challenges to shooting the same scene over and over or performing a play dozens or hundreds of times. Some theories suggest that the nature of this process means that what you see in this predictable and repetitive environment of performance might be more a combination of what is really there and what you expect to be there from previous experiences and predispositions.[4] We know that our brains use attention to detect what is *different* than what we expect in a moment or place or interaction.[5] That is incredibly helpful for you as an actor. Each take or performance inevitably generates some differences in your partner. To retain spontaneity in your performance, look for difference and use that to fuel your response on that day and in that specific moment.

If your partner plays "flicks" in the same spot every night and, as they say the line, they turn away from you toward their desk – how is it different tonight? What is new about that moment? Do they turn at exactly the same time? How do they initiate? How is their voice different today? Can you find a new place in your own body where you experience being flicked? Keep your task outside of yourself and on your partner. *Don't just see them. Look for difference.* Then you can resist some of that natural inclination to check out and let the simulation of this moment in your brain do all the work.

Perhaps your partner is willing to play with some of the earlier exercises as a warm-up for performance. If not, warm-up on your own by expanding some of the impulses. Check in on your Action Palette Diagram. Has it shifted since you began? Are there actions that you forgot that would help deepen a relationship? Try rereading your notes from the first read. How many of your own questions and expectations are alive in your performance, and to what extent? Can you tell from the audience response if they're taking the same journey you did when first reading the script? Returning to some of your own earlier work illuminates what changed and invites you to rediscover what slipped away through habit or repetition. The physical actions and diagrams and palettes may not need to be on your mind moment-to-moment, but they become a resource. The tools of attention and embodiment must be a constant part of your dialogue with yourself about your daily reinvestment in your work.

Notes

1. D'Angelo, Mariano, Giuseppe di Pellegrino, Stefano Seriani, Paolo Gallina, and Francesca Frassinetti. "The Sense of Agency Shapes Body Schema and Peripersonal Space." *Nature: Scientific Reports* 8 (2018). DOI: 10.1038/s41598–018–32238-z.
2. May, Michael. Personal communication via Facebook Messenger, February 24, 2019.
3. Sutton, John and Kellie Williamson. "Embodied Remembering." In *The Routledge Handbook of Embodied Cognition*, edited by Lawrence Shapiro. New York: Routledge, 2014, p. 316.
4. Clark, Andy. *Surfing Uncertainty: Prediction, Action, and the Embodied Mind.* Oxford: Oxford University Press, 2015, pp. 13–14. DOI: 10.1093/acprof:oso/9780190217013.001.0001.
5. Ibid., pp. 58–59.

14 Skills Gymnasium

Doing Reps

You may find that stepping away from work on a particular role or script helps you better understand physical actions and build skills to bring back to text work. This chapter includes a number of experiments that you can use in a class or ensemble setting unrelated to work on a particular scene or project. They build certain muscles and allow you the chance to focus more closely on those concepts. First, let's return to observation of the world. This experiment is best done in a public space where people tend to stop for a while rather than simply walk past. Coffee shops, parks, hospital waiting rooms, and bus or train terminals are all good options.

Observation of Public Space

In pairs, find a spot in a public space where you can sit for a while, observe, and listen to others. Try your best not to be so close to them that you change the dynamics or so far from them that you have to completely guess at conversations or behaviors. Have a notebook ready.

1. *Sit for about 30 minutes. Be there long enough that you'll see and hear a variety of interactions from different people.* Maybe set a timer on a device so you don't keep checking for how long it has been. Don't use the device while you observe. It will alter what you manage to see and hear.
2. *As you observe various events, try to name some of the actions people are playing on one another.* What kinds of actions do people play to start interactions? To end them?
3. *Don't just listen for how they play physical actions through speech, try to observe all of the ways that their actions are embodied in that moment.* What is the angle of their body? Where do their eyes look? Are they engaged in an activity?
4. *As you observe, write down the physical actions you think you can name in the order they occur.* When you can't quite figure it

out, simply write what was said or a few words to help you remember the behavior you observed. It is like a Score of Actions for this real-life scene.

5 *You don't have to write down every single thing you observe.* Some events might be too brief or simply not grab your attention. Focus on the ones that are most interesting to you.
6 *At the end of the 30 minutes, find a spot to regroup with your partner and compare notes.* Did they come up with similar actions? Did they find the same events interesting? How was their take different from yours?
7 *Back in the studio, take one of those scenes you observed (perhaps a 30–60 second event) and re-create it action-by-action.* Use additional class members if there were more than two people involved. Take some time to do this. Rehearse it like a scene with specific text and blocking.
8 *While the rest of the class watches, perform the scene you observed.* Try to replicate what you saw as specifically as possible.
9 *As the rest of the class watches, they should write down every action they can name.*
10 *Perform it a couple of times so they can refine their observations.*
11 *After you're done, hear some actions they observed. Don't tell them what you actually wrote at first.* How close were they to what you tried to play? Did they all misread your actions in the same way?
12 *Try a few moments again, refining the actions to help ensure they see it more clearly.* Maybe expand the physical action some. Or strengthen the way it comes out on language. Does this take you further from what you observed? Closer? How do those same actions read differently on you than on the person you saw? What adjustments are needed to make the same actions clear in your own body?

There is a risk that in replicating these (likely) mundane events you will try to make them more theatrical. If the person dropped their wallet and then picked it up, no need to invent more dismay or panic than you actually saw. Resist that impulse. You must trust that the detail and specificity with which the people you observed played physical actions is sufficiently interesting. Working this way is a little like a game of physical action "telephone." Some translation is inevitable, but your classmates are a valuable resource for how you are, and are not, communicating the actions.

Embodied Guessing

Along these same lines, it helps to explore how much must be visible for others to see or understand an action you're playing. A physical action can feel incredibly strong when that internal dial is up to 10, but it may not always be so clear to an audience. Try this next experiment to practice the level of visible physical engagement required to make different actions clear.

Expand and Contract – What's My Action?

1. *Partner A selects a physical action or is given one by the teacher. Partner B joins them to be the receiver.*
2. *Without speaking, Partner A begins to find the smallest physical form of that action they can play.* Perhaps it is just a look, or a gesture, or behavior.
3. *Partner B responds with whatever they think makes sense based on the action they are getting.*
4. *The audience begins to guess out loud (like in a game of charades) what the action is.* Don't stop if only one person guesses it correctly. Wait for there to be group consensus.
5. *Keep expanding the action and clarifying.*
6. *Use Partner B as a guide for both Partner A and the audience.* Are they responding in a way that seems to understand your action? If not, adjust and expand to find specificity.
7. *When a majority of the audience is guessing the correct action (or a synonym) you can stop.*
8. *How expanded did the action need to be before they got it?* How did the receiver help or hurt their ability to guess correctly?
9. *Try the original version again. Can you make an adjustment to keep it this contracted and yet clearer than it was at first?*

It isn't necessary that an audience understands every physical action that is being played at every moment. Sometimes the action is quite purposely masked from the other character and the audience because of the circumstances. It is important, however, that the actor is playing something clearly and specifically and using the tools of embodiment to achieve that goal.

The Resistance

Some actors struggle to understand how important it is to give their partner something to struggle against. Either that, or they simply give their partner back the exact same action they are getting, locking you into an endless loop of yelling or imitation. Neither of these makes for particularly interesting work. Good conflict requires good resistance. It also requires choices that provide a path to something new on the other side. Try the following experiment to illustrate this.

Push, Pull, Dodge

1. *Two actors face one another. One is Partner A, the other is Partner B.*
2. *Both place their hands out in front of them, palms facing their partner. Touch palms.*
3. *Partner A pushes. Partner B pushes back.* Notice how you work to win. Are you evenly matched? How does each partner compensate for the power of the other?
4. *Now, Partner A pushes and Partner B pulls.* See how quickly it ends. Who seems like they won? Did Partner B really pull or just let the energy of Partner A do the work for them?
5. *Finally, Partner A pushes and Partner B plays "dodges" or "avoids." How* does Partner A solve the problem of Partner B's new action?

In this exercise you quickly discover how important it is to have two strong and complimentary actions. In the push/push version, this stalemate could go on endlessly – the scene never making progress. If you constantly match your partner, you add very little information and provide few paths for the scene's journey. In the push/pull version you see how rapidly a scene is over when your action simply helps the other person achieve their goal. Not only is there nothing to push against, but you achieve your own goal with almost no effort on your part. In the final version you see how much opportunity arises from two strong well-matched actions. Here Partner A must constantly reassess and calibrate to an ever-changing partner. Partner B is active and responsive and resisting Partner A, but also has their own specific goal. The nature of the action itself is in constant transformation because of your partner.

Cascades

Initial actions often fail, and you must either escalate or de-escalate in order to achieve your goal. These cascades of actions appear in texts, but it is helpful to explore them on their own, too. As a reminder, these are series of actions that all happen, one after the other, toward a specific goal. They often result from a failure of each action to get the desired result from a partner, thereby forcing the shift.

Cascades Gymnasium

1. *Write several cascades of actions on slips of paper. Select 3–4 actions for each cascade.* Make sure these lists are not all the same kinds of actions or always cascading in the same direction. For instance, try "pokes, pushes, shoves" or "worships, adores,

admires." The first example escalates and are all primary actions. The second set de-escalates and are conceptual actions. You can also mix and match as in "kindles, rallies, shakes," which moves from metaphorical to conceptual to primary as it builds.

2 *Decide who is Partner A and who is B.*
3 *Begin in thick space with very expanded actions. Repeat the cascade a number of times, contracting the visible form and allowing the space to shift with each repetition.*
4 *Partner A plays the cascade, allowing Partner B to respond to each action before moving on.* Make sure it is still a conversation and not just a monologue.
5 *For Partner A, keep refining what is most effective for each action and what about their response makes you want to escalate or de-escalate?*
6 *For Partner B, what actions can you play in response that most feed the cascade? How can you offer resistance with your action selection and help strengthen your partner's impulse for the next thing?*
7 *After several repetitions, reverse who is A and who is B.* How is it different? Don't simply replay what your partner did. Discover this as an entirely new version.
8 *Be sure to experiment with different levels of expansion and contraction.*
9 *Don't forget language.* Try saying the action or finding words that help support and strengthen the impulse.

Many students discover that these cascades really do require harmony between the partners. The score of actions and their build doesn't make sense unless the right kind of resistance comes from a partner. This is certainly true in scripted text, as well. While action selection can be a very personal process, the actions must serve the story being told. If it is clear from the text (or the director) that a cascade is necessary, each partner should seek to craft a kind of opposite cascade of responding and resisting actions. Without that harmony, Partner A seems to be pushing against nothing at all.

Individuality

As soon as someone new plays a physical action the very nature of the exchange transforms. There's a great deal we share about how we think about these words, but also many ways they are utterly personal. This next experiment looks at these differences and helps expand your thinking about embodying actions in new ways.

One Action, Many Versions

1. *One person stands in the center of the space with several others in a large circle around them.*
2. *The person in the middle selects a physical action (or assign one to them).*
3. *Starting with a fully expanded embodiment, they play that action several times with one of the people on the outside.* The person on the outside responds with whatever action they'd like.
4. *Keep refining that action to impact them more clearly.* Resist changing the action and turning it into a cascade through repetition.
5. *After a few attempts, move on to the next person. Really start all over. Experience the different way this person is responding.* How does that change the way you play the action? Maybe you need to contract or even expand further to be more effective. Maybe a different part of your body needs to begin the impulse. Maybe you need language.
6. *Keep moving around the circle like this. Allow your focus to be "How do I play this action with this specific person?"* It isn't about your sense of the action, it's about how it lands on and impacts your partner.
7. *After this person has gone around the entire circle, have the final person they played with come to the center and begin to play that same action, working their way around the circle.* How is the same action different in this new body? How do the same people respond differently to them?
8. *You can keep going like this with the same action or pick a new action after a few people have gone around the circle.*

This experiment illuminates just how much you slip back into making the physical actions about yourself or your experience of playing them. If you simply repeat the same physicalization or language with each new person then you aren't really taking them in. It's easier to simply send the action out into the world and move on. By repeating it with different partners, you discover how important it is to follow up the action with immediate curiosity and attention. How will you know what comes next if you don't see what comes back?

Triangles

Some actors become too focused on the two-person nature of playing physical actions. While most of the time an action is aimed at a single person, sometimes the target is two or more people who serve as a single entity for a moment. It is also true that with more people on stage you may experience that the target of your actions jumps around moment-to-moment and line-to-line. It's valuable to experiment a bit with how you make those adjustments, frequently without even realizing you are doing it. A simple example might be in a classroom. Sometimes a student provides feedback to another student. Certainly, there are times the action is being played on the other student, but there are also times when the action is being played on the teacher or someone else in the group. In a way, this is like our experiments with objects and furniture. The person you're speaking to becomes the object that you bounce the action off in your attempt to land it on someone else. It might feel unkind to do this, but it can be an innocuously unconscious act or an intentionally cruel one depending on the circumstances.

Triangles

1. *Three actors stand in a triangle, giving lots of space between them. Select who is A, B, and C.*
2. *Write a few dozen physical actions on pieces of paper and put them in a container.*
3. *Partner A pulls out an action and immediately tries to play it on B and C at the same time.* How do they react differently? Does one require more attention that the other? How do you track the collective and individual success or failure?
4. *Partner A pulls a second action from the container. They continue to play the first action with one partner and the new action with the other.* How do you switch back and forth? Why did you stick with the first action with that partner? Are you spending even amounts of time with each?
5. *As Partner A works, B and C respond with whatever actions arise for them.* You do not have to keep repeating the same action. Perhaps the repetition from Partner A generates new impulses. Maybe their split focus on you and the other person creates an impulse for new actions.
6. *After several repetitions, Partner A should switch which action they play with which partner.* What is the new dynamic given the difference in their responses?
7. *Once you work in this way, have Partner B take the lead and repeat the exercise. Then Partner C.*

Variations: While it is often easiest to see the impact working with the most expanded versions of the actions, try a variety of levels of visible embodiment as you work. Are you using language? Saying the action out loud? Adjusting all of these might be more, or less, effective depending on the partner. Don't assume the same forms of embodiment will work with everyone or with every action for you.

Sometimes this experiment with triangles generates an exciting sense of relationships. It is a micro version of switching palettes when two people don't overlap much in your Action Palette Diagram. You can also select a relationship in advance to help fuel the stakes of the physical actions. A parent and two children or three co-workers or romantic partners and one of their roommates all produce useful dynamics. The risk of these scenarios is that you begin to rely on language and plot in a way that avoids a focus on action and embodiment. You can also simply select a status (between 1 and 10) for each person and see the ways that plays out in a three-person dynamic. A lower-status partner playing "destroys" to both a higher-status partner and another low-status partner, for instance, cracks open discoveries about how many ways a single action manifests itself. Experiment a bit further with status now.

Status

1 *Working in pairs, have one person select a physical action and a status between 1 (low) and 10 (high). The other partner simply selects a status.*
2 *Experiment, through several repetitions, how this action operates with this status.* Does it change how expanded the embodiment can be? Does it make spoken language more or less useful? Does it help to initiate from a particular part of your body?
3 *The other partner responds with whatever actions arise from the action they receive filtered through their own status.*
4 *Keep refining and experimenting.*
5 *After several repetitions, Partner A should keep the same action but switch status.* The shift need not always be massive. Sometimes experiment with just altering it by a point or two. Play with being close to or the same status as your partner. Discover the ways your sense of status shapes and squeezes how you play each action by changing the rules of the world.
6 *Switch back and forth with who is leading with a specific action and who is simply choosing a status.*

You can do this with several pairs working at once or with the rest of the group watching. It's helpful to have someone else call out the change in status for each partner. Be careful not to change status too frequently. It takes away the opportunity to refine and explore the different ways to become effective at landing the action. Reflect on your idea of status as you work. Do you have a set of habits or personal or cultural markers that define it? Do you and your partner share those? In life, status is either an agreement or imposed from outside. How would your actions change depending on which version this is?

What is "Realistic"?

One of the challenges of exploring embodiment is our own sense of how physically engaged we are in the world. Some of the more expanded explorations of physical actions might feel incredibly unfamiliar or challenge a sense of truthfulness or realism. Some of these factors are personal, of course. Families have their own spoken and unspoken rules about how loud you speak or how you hold yourself physically in public. There are also larger community and societal expectations that are either shared or applied differently depending on group identities. All of this complicates embodiment in performance and especially how you experience bigger events in the life of a role. Your own life may contain many high-stakes and profoundly dramatic moments. In all likelihood though, these are brief and rare and maybe even traumatic and so your particular memory of the embodiment is limited.

It's helpful to do the following exercise to explore that idea of "truthfulness" in more extreme moments. If you can, use a piece of text from a script. If you don't have one, it is fine to invent a sentence or two and create a scenario. I don't find it useful to pick a real thing that happened to you or actual words you spoke. Trust your imaginative engagement.

Electric Fence

1 *Working with a partner, speak the text and allow the rest of your body to respond in whatever feels like a "truthful" manner.* Really try to land the text on your partner. How do you want to change them? What, if any, impact does it seem to have on them?
2 *Repeat the text again and allow those physical impulses to expand a bit. It is a little like Expand and Contract, but let the expansion be tiny, don't move into thick space or abstract the physicality from the world you've imagined.*
3 *Continue doing this. Bit by bit. Let the repetition and the additional freedom strengthen the sense of the action inside you.* It is okay to make non-word sounds if it helps. Can you turn up that internal dial?
4 *Keep the stakes high.* Make it really important that this changes the other person.
5 *Keep going like this until you feel you hit the "electric fence" – that moment where if you go any bigger it will feel fake or too big.*
6 *When you feel that, do it one more time, just a little more expanded.*
7 *Were you right? Did you still find a truthful version? If so, go a little further.* See if you can expand your idea of truthful, not by simply being bigger, but by using the expansion to support a stronger and stronger action that *demands* a bigger expression.

> *Variation*: Try starting the same way but getting smaller and smaller in the embodiment. How small can you get while maintaining the strength of impulse and need before you hit the fence?

This experiment is a helpful diagnostic tool. Allowing incremental shifts in the level of visible embodiment helps actors develop a way to live truthfully in bigger and smaller ways. I think of it as a fence because I so often see actors slow down as they get closer and closer to it. There is some sense that they're approaching the edge of something unpleasant or that something like a force field is just beyond. It's often necessary to go right up to that edge and then play back and forth beyond it. Experimenting in that space sometimes opens up a new frontier in how expanded an action can be while staying truthful. You discover that the fence is actually much further away than you realized. For some actors, it is an important experiment to help recalibrate their sense of truth, especially if they work to keep things small and contained all the time.

I have no doubt that these experiments will spark variations and discoveries and adjustments in their form as your own project or class dictates. Importantly, you must stay rigorous about the core function of each exercise. If your action, no matter how expanded or contracted, is just a *show* for your partner rather than a direct attempt to alter them *with* it then you might as well be alone. Every word, inflection, gesture, move, use of an object, and breath is a part of your embodied attempt to impact the person across from you. Anything extraneous is about you. A body deeply engaged in trying to alter the other person is a body full of vital information for the audience to lean forward and watch closely. A good director can help you know the difference, but your best resource is your partner. Are they changed? Truly? If not, how can you adjust to be more impactful within the Rules of this World?

Many Tools, But One Goal

In Part 4 you explored how to experiment with physical actioning in a variety of ways with partners and ensembles. This included exploring the way your Primary Action Palette and Relationship Palettes come into conflict or harmony with others. You also explored the relationship history and major events through embodied experiments. You discovered how stretching out time with thick space and the use of music enhances links to the events and actions. You returned to a careful examination of text to see how all of this embodied exploration enriches your understanding of the moment-to-moment work of living inside the words. Finally, you discovered a variety of experiments away from text to deepen your skills and refine your understanding of how physical actioning works in your own body and between you and others.

If you began all the way back in Part 1 with the earliest experiments in curiosity and attention and followed these explorations through to now, both on your own and with a class, you have a comprehensive set of tools to apply in nearly any circumstances where you are called on to represent or recreate a sense of humanness. As you reflect on all of these experiments, terms, and ideas, keep your north star the simple goal of using action to change your partner and to stay curious and attentive to the result of those actions. Everything else is a way to hone and refine and deepen your capacity to do that simple but elusive job.

This book isn't an instruction manual for how to act. There are, inevitably, aspects of the work that go unexplored here. It is also important to remember that no single set of tools works for everyone or all the time for anyone. More than anything, I want you to recognize that your work and the way you engage with it is an inherently embodied process. How wonderful to be living and breathing and three-dimensional in space and time. How exciting to be working and learning at a moment when we understand that the very thought of movement is, in some ways, movement itself.

Release your old and well-established ideas about how an actor prepares a role and the order and structure of that preparation. Can you imagine acknowledging and honoring and engaging your whole body in your work from the beginning? Can you imagine embracing new information about how human beings work to better use yourself to communicate and impact your audiences? That is what I hope these tools offer you, both as you walk through the world and the next time you pick up a script – a journey full of curiosity, turned to attention and, then, action.

Part V

Program Notes

15 Would Stanislavsky Disapprove?

Zooming Back Out

If you read Part 1 through Part 4 you have all the tools you need to explore on your own and with collaborators while ensuring that your work stays embodied from the first moment right through to performance. As you apply these techniques, you'll certainly collaborate with people who work in other ways or you'll take classes that seem to offer different points of view. Learning a variety of sometimes seemingly contradictory tools can be incredibly valuable. It's good to grapple with what does and doesn't work for you. It's helpful to discover that some tools fit certain jobs or challenges while others serve you better over the arc of a project. Still, it can be confusing to know how to blend ideas or when to reach for which tool in the toolbox.

When I teach this work, two big questions come up about the relationship of physical actioning to our profession and other actor training tools: "Does this work 'reject' Stanislavsky's ideas and methods?" and "If I understand what's happening in my brain will it mess up my acting?" The short answer to both is "No," but a longer answer would probably help. Part 5 is the longer answer to those questions. Think of this final section of the book like program notes for a play. If you're interested in this way of working or are teaching with these experiments, it will help you place physical actioning in relation to other acting tools and techniques and dives a little deeper into the key ideas about neuroscience that are sprinkled through the rest of the book. Much like the program notes for a play, you can get a lot out of the work without reading them, yet reading them enriches and deepens your understanding of the whole. I hope you'll find this last section does the same.

Stanislavsky and the Mind

Sometimes when I mention neuroscience to actors and fellow teachers their response looks a bit like someone hitting a wall of unpleasant smell upon entering a room. I understand. The thought of integrating theories developed in labs and with giant fMRI machines feels antithetical to the impulse-based individualized creativity that so often draws people to become theatre-makers. What they forget is that curiosity about modern science and scientifically investigating humanness is already at the heart of what we do. Most actor training in the United States is still based on the work of Stanislavsky, a branch that grew from that tree, or work built on his ideas. The physical actioning described in this book, however interested in incorporating newer knowledge and theories about embodiment, certainly shares that same set of values, history, and vocabulary about behaving truthfully and in a lifelike way.

You may take for granted the relationship between acting and psychology because psychology is so much a part of how we speak about the work and it is assumed to be necessary to understand the way characters behave. It is easy to forget that Stanislavsky was doing something pretty new and, importantly, was curious about both the emerging science of his time as well as the larger shift toward understanding humanness through a psychological lens.[1] He specifically references the translated writing of French psychologist Théodule Ribot. As Stanislavsky's ideas developed, he became more focused on the physical elements of the actor's work rather than generating emotion or psychology itself as a starting point.

Late in his career, Stanislavsky developed the "Method of Physical Action" to explore some of his new ideas about beginning from the physical rather than the emotional or psychological. An important element of this was his introduction to the ideas of William James and Carl Lange who theorized that emotions began as a physiological response that was then experienced as emotion when those physical signals were received by the brain.[2, 3] Essentially, that means that James and Lange believed that the physical sensation was first and that, once your brain *registered* that sensation, you understood you were experiencing an emotion. This was further influenced by additional work on the relationship between behavior and thought/emotional response. Some people link Stanislavsky and Freud. The evidence doesn't indicate that Stanislavsky knew or engaged with Freud's ideas directly and, given that, Stanislavsky's integration of psychological principles lacks concepts that developed later, and even some that existed during Stanislavsky's life.[4] These later explorations about physical action were never fully included by many acting schools and methods rooted in Stanislavsky's earlier work.

The modern impulse to embrace what neuroscience can teach us about acting is the same one Stanislavsky had about psychology – he wanted to understand how humans work to build better tools to replicate that humanness on stage. Not only that, he wanted us to build on his ideas. As Jean Benedetti writes in his introduction to *Stanislavski and the Actor*,

> Stanislavski always insisted that his work had to be useful and that it should be extended and developed … The one condition he laid down was that its basic principles, which he believed to be rooted in human biology, should be respected.[5]

Someone so curious about the advanced ideas of his own time would almost certainly be curious about the new knowledge we possess concerning the very questions he was asking.

If you know about the work Stanislavsky did later in his life, then you know the term "physical action" itself is not new. Much has been written about the way Stanislavsky's work was misunderstood, misappropriated, and mistranslated in the journey from Russia to Europe and, particularly, the United States. His later work on the Method of Physical Action was incomplete by the end of his life – though even the idea of "complete" might have bothered him. He resisted the impulse to call some segment of the work definitive.[6]

The Origin of Impulse

In reading about Stanislavsky's later work, you see him grappling with this question of the origin of impulses. He recognizes that the way impulses manifest in bodies on stage is integrated with thought. He is, however, using the knowledge of the time and many of the exercises and ideas focus on one element or the other rather than seeing the mind as a fully integrated part of the body. Working this way sets the stage for many of the conversations actors and teachers continue to have about "working from the inside out or the outside in." The directionality in these conversations is built, in part, on this idea that the expressiveness of the body is in the service of allowing an audience to see the psychological.[7] It is hard for us, even today, to conceive of the totality of our systems and the way each part affects and alters the others.

Stanislavsky uses this duality when discussing "Mental Action" and "Organic Actions." The first focused on the imaginative or emotional and the later on physical activities or behaviors with more, or less, complex psychological underpinnings. Certain tasks, such as cooking a meal, might be an Organic Action, intended to show affection for a partner or to bring to a neighbor who had lost their home in a fire. It is the same task, but the psychological content is different depending on those circumstances.[8]

Even his misunderstood work on "Emotion Memory" included the exploration of physical engagement to spark psychological response. He had actors recall emotional events and then invent a circumstance for the exercise that might generate a similar response.[9] This is not an abandonment of the imaginary circumstances and a retreat into personal memory, as actors sometimes misunderstand it to be. Rather than simply thinking about a similar real event, Stanislavsky asked the actor to embody a situation that might provoke a parallel response. He knew that small physical triggers and impulses could be powerful for the actor and trusted that these physical experiences would be sufficient for a well-trained actor to have an emotional response.[10]

Stanislavsky recognized that a body able to fully and freely express is a core requirement of the actor. He also demonstrated a clear understanding that experiencing physical events and moments are an important way to achieve the "result" of emotional response. He acknowledged that emotion is not generated merely by thinking about the emotion itself, but rather by the use of imaginative engagement with events or circumstances that are likely to produce the emotional response.[11] He saw that *doing* was a path to feeling.

The Vocabulary of the Moment

In each of these examples, Stanislavsky embraces the physical as a path to the psychological, and vice versa, but understandably, given the knowledge of the human mind at the time, maintains a dualism in the approach. This is difficult to overcome. Any exercise requires you, as an actor, to put your attention on some element of the whole. The challenge you now confront as a modern actor, is how to use the complexity of your physical system and unify the way you speak about your impulses to embrace that reality. That is what the tools in this book try to help you accomplish. Because of his moment in history, some areas of Stanislavsky's work make distinctions that are less accurate or imagine a path of practice that doesn't mirror the way we now know humans learn.

One example of this is in Stanislavsky's three "Modes of Communication." The first one is verbal (the examples of which would be obvious), the next one is gestural, and the final one is mental, where he describes a kind of telepathy of sending and receiving thoughts and feelings non-verbally.[12] In this we see verbal expression treated in a hierarchy above gesture for communication. As many of the exercises throughout this book illuminate and, as you'll discover in more detail in Chapter 16, we now know that these two are deeply connected and do not necessarily exist in a hierarchical structure.

His Method of Physical Action also included a process of working with text that began seated at a table and eventually allowed for a more and more embodied experience. This is a kind of supremacy of written text and verbal expression that still plays out in most modern rehearsal rooms. There is an underlying assumption that the physical is best discovered as an outgrowth of a distinct analytical process. In describing the stepped approach to physical engagement with a script his method starts with embodied exercises, but abandons it when the text comes into play. The process involves reading the text aloud (while sitting on your hands), adding gestural movement while seated, then improvised behavior while standing, and eventually focusing on creating character by more deliberate choices.[13] We now have a clearer understanding of how our brains process written text differently from speech and that speech has other inherently physical impulses associated. A more embodied approach with text from the beginning could achieve the connections he desires and remove the hierarchy of speech.

Same Vocabulary, Different Meanings

One additional distinction about how Stanislavsky thought about physical action and how you explored it in this book is about what constitutes an "action." When Stanislavsky spoke or wrote about actions, he was mostly describing what I call "activities" and "behavior." This might include elements like blocking or handling of objects or physical interactions with other actors. As you now know, the underlying impulse (which I call the "physical action") happens in a much broader set of ways – from gesture to speech to breath shifts to your palms sweating and on and on. That's not to say that he didn't realize the behaviors were related to thought or driven by deeper needs, just that he often described a narrower set of possible ways thought manifests in the rest of the body.[14]

I offer this background partly to give credit where it is due, but also because actor training, particularly in the United States, is still tightly focused on a narrow set of techniques that share some of this limited thinking about embodiment. A careful reading of history and an understanding of Stanislavsky's own journey reveals how unnecessary that is. Any set of tools or techniques that dismisses our scientific progress in understanding the mind ignores the core values that lead Stanislavsky's own explorations. Perhaps the acting teacher most responsible for the next leap forward in unifying the mind and body in actor training is also most often mistakenly used as a contrast to Stanislavsky: Jerzy Grotowski.

Grotowski's Expansion of the Physical

While many people think of the work of Polish director and teacher Jerzy Grotowski as totally different from, or even anti-, Stanislavsky, he described it as a continuation from where Stanislavsky left off. In fact, he carried the term "physical action" itself forward. Remarkably, many teachers and schools still see their work as opposite poles or even contradictory methods. Thomas Richards, a long-time student and collaborator of Grotowski and author of *At Work with Grotowski on Physical Actions* quotes Grotowski both crediting Stanislavsky as the basis of his own work as well as being a "fanatic."[15] It is clear from Richards' own experience of Grotowski that this shows up in a particular and specific passion for Stanislavsky's unfinished work on the Method of Physical Actions.[16]

Grotowski expanded on the tools and did a lot to build past the way Stanislavsky's exercises focused on "activity" as action. He worked toward a broader idea of embodiment and expression as an engine for a psychological and emotional response. In talking about where physical actions originate in people, Richards quotes him speaking at a conference in Liège, Belgium, in 1986 describing the origin of physical actions saying "The impulses: it is as if the physical action, still almost invisible, was already born in the body."[17] As an example, Richards describes his first encounter with Grotowski's work through long-time company member Ryszard Cieslak. He writes about watching Cieslak cry like a child and how clearly his method was focused on reproducing the child's actions rather than the feelings that might produce crying as a product.[18]

The focus, however, was not merely the embodiment of emotion. Many of his exercises, like his "plastiques," develop a labile and expressive body and rigorously prepare it to respond to a scene and partner. The mistake many observers (and even students) make is to assume that the "unrealistic" expressiveness of any given exercise is intended to communicate the values every final performance should possess or is a statement of performance aesthetics. As you've seen in this book, there are many ways to find that expanded expression of the impulse and then harness and shape it to fit the rules of any world. The final performance doesn't need to look like an abstract movement piece to benefit from expansive explorations with the whole body during the process.

Activities, Movements, and Gestures

At the same time, Grotowski made certain that people didn't oversimplify the way the body should be engaged in performance. In speaking about what physical actions are *not*, he helpfully describes three categories that many students fall into when they first begin to explore the concept. These are activities, gestures, and movements (or, as I might call it, blocking). Combined, they add up to what we refer to throughout this book as activities and behavior.

Grotowski wisely argues that activities, such as sewing a button, only become physical actions when they occur in response to or in relationship with the partner. To sew alone is not necessarily physical action. If you're sewing while your partner is on the phone with their ex, perhaps that sewing becomes a physical action through that circumstance.[19] Remember our explorations bouncing a physical action off of an object to a partner or using an activity to play actions on your partner. This is very much in keeping with how Grotowski describes the need to link activity to partner for it to truly become an action and even related to Stanislavsky's idea of how Organic Actions relate to the psychological.

Similar to activities, he argues that movements (or blocking), like sitting, walking, or exiting are not inherently physical actions – they only become so when they are motivated by the response to a partner. If you stand from sitting to move away from an acquaintance who always asks about your vegetarianism, standing has become a physical action.[20] In these descriptions, Grotowski clarifies that any visible behavior a person does is not an action merely by being visible. It becomes an action only if there is a need, a "why," underpinning it. This cannot be a merely technical "why," as in "I am leaving the room so I go to the door." It must be in relation to a partner and be directed at them in some way.[21] In his discussion of movements Grotowski describes a "cycle of little actions." This is an example of your "Score of Actions, or, as he called it "individual structure."[22] He saw this as a building block of sorts.

Grotowski seems to argue that gestures are not physical actions because they are "not born from the inside of the body, but from the periphery (the hands and the face)."[23] The emerging science indicates he was wrong on this point. There is a direct link between organic gesture and thought. Perhaps he's responding to a kind of performance where empty or planned gesture serves as a replacement for real action. That, of course, is not physical action. True impulse-born gesture, however, is absolutely a manifestation of and a part of the way you play physical actions. Some of these instructions are admonitions about the aesthetics of performance from the time. Certainly, however, actors still engage in unmotivated behaviors that lack a connection to partner or a story-driven goal.

In life, *all* activities, behavior, gesture, and movement are goal-driven. These moment-to-moment actions contribute to and are caused by larger needs. Our work together in this book helps you understand how those goals manifest themselves, even in very small ways, throughout the body at all times. Then you can replicate that process in crafting a performance within imaginary circumstances.

The work Grotowski did to further explore ideas around embodiment and to deepen the work of physical actions beyond the "realistic" were prescient about the way the mind operates as part of the body. Still, he did not have the benefit of modern neuroscience to link the elements completely. Regrettably, rather than further integrating the ideas of Stanislavsky and Grotowski, many actor training programs selected one or the other as primary, considering it an aesthetic decision between a realistic theatre of psychology or a "physical theatre." This only deepened misguided thinking about mind/body dualism.

You will come in contact with many fellow actors and directors who (knowingly or not) hold these beliefs or use the language of these methods in the way described above. Understanding how these foundational techniques serve to support your own exploration of new integrated tools is valuable as you collaborate. It doesn't require you to disagree or challenge collaborators, but it does allow you to hear their language and translate it for yourself into a more holistic approach that encourages your full self to participate from the start and throughout the process.

Summary

While most modern actor training techniques are branches of the tree of Stanislavsky's work from about a century ago, some errors in understanding and conflicts between teachers still prevent us from seeing both the shared origins of techniques and underlying assumptions about how human beings work. This keeps us from embracing new discoveries. Stanislavsky was curious and engaged in using emerging science about human psychology to develop better exercises and techniques. The last years of his life included the development of the "Method of Physical Action," which moved toward physical engagement with the space and the physicalization of imagined events as a core starting point for "truthful" acting. Importantly, he believed that people should continue to develop his techniques and that the techniques would inevitably require adjustments both because of unique cultural and national demands on actors and as a result of gaining a better understanding of the nature of being human in the future.

While many students and teachers in the United States and elsewhere did not embrace these later developments in his work, Polish director and teacher Jerzy Grotowski used his training in Stanislavsky's work to expand the Method of Physical Actions even further. He developed his own ideas about the way human beings physicalize thoughts and proved impressively prescient about later discoveries in neuroscience. While many people see a split between an "inside out" approach by Stanislavsky and an "outside in" approach by Grotowski, their own work suggests that a hard break is false. The idea is more a reflection of our own historic thinking about a divide between the mind and the body than the work of these particular teachers. This bifurcation has continued for decades, with various teachers and methods attempting to build bridges between them.

We stand at an exciting moment, however, when our rapidly expanding understanding of the brain allows us to not only test the tools developed by Stanislavsky, Grotowski, and others, but also liberates us to develop new tools that eliminate some of the assumptions about the mind/body separation they were built on. So, "Would Stanislavsky disapprove?" Rest easy. I firmly believe that if they were alive today, both Stanislavsky and Grotowski would be voraciously curious explorers of cognitive neuroscience.

Notes

1 Blair, Rhonda. *The Actor, Image, and Action: Acting and Cognitive Neuroscience*. New York: Routledge, 2008. 30.
2 Ibid., p. 31.
3 Cannon, Walter B. "The James-Lange Theory of Emotions: A Critical Examination and an Alternative Theory." *The American Journal of Psychology* 100, no. 3/4 (1987): 567–586. https://www.jstor.org/stable/1422695.
4 Sullivan, John J. "Stanislavski and Freud." *Tulane Drama Review* 9, no. 1 (1964): 88–111. https://www.jstor.org/stable/1124782.
5 Benedetti, Jean. *Stanislavski and the Actor*. New York: Routledge, 1998, p. xiv.
6 Ibid., p. 104.
7 Ibid., p. 13.
8 Ibid., pp. 27–28.
9 Ibid., pp. 66–68.
10 Benedetti, Jean. *Stanislavski: An Introduction*. New York: Routledge, 2004, p. 92.
11 Ibid., pp. 92–93.
12 Benedetti, Jean. *Stanislavski and the Actor*. New York: Routledge, 1998, pp. 76–79.
13 Ibid., p. 107.
14 Ibid., p. 26.
15 Richards, Thomas. *At Work with Grotowski on Physical Actions*. New York: Routledge, 1995, p. 6.
16 Ibid., pp. 4–5.
17 Ibid., p. 94.
18 Ibid., pp. 12–13.
19 Ibid., pp. 74–75.
20 Ibid., p. 76.

21 Kemp, Rick. *Embodied Acting: What Neuroscience Tells Us about Performance*. New York: Routledge, 2012, p. 181.
22 Richards, Thomas. *At Work with Grotowski on Physical Actions*. New York: Routledge, 1995, p. 65.
23 Ibid., pp. 74–75.

16 Will My Brain Get in My Way?

We live at an exciting moment when scientists are deeply engaged in asking questions about humanness similar to those asked by actors and theatre-makers for centuries. In addition, theatre scholars are linking those investigations with our work. You don't need to be a scientist or theatre scholar to learn from and appreciate how their work helps your training and in performance. Many of the excellent sources referenced throughout this book dig deeply into neuroscience and the relationship to acting, theatre, and audiences. I largely focused on tools that you can practically apply – hoping to build upon that theoretical work. It is important to remember that these ideas about the brain are still developing and it's a complex challenge to apply them outside a scientific laboratory to the even more variable laboratory of acting. The work of physical actioning is something you can use whether or not you fully understand the science and theories behind the work. It is practical in application even when it is experimental in its ideas. As our understanding of the brain develops, exciting new ways to shift and refine the exercises will also emerge.

If you're like my students, after asking "Would Stanislavsky disapprove?" you might then ask that second big question, "If I understand what's happening in my brain will it mess up my acting?" Some people resist learning these ideas early on because they fear it will rob them of a sense of spontaneity or something mysterious about acting. There is a superstition and mythology about innate talent and impulsiveness that surrounds the stories we often hear about great actors and how they work. While some small fraction of excellent actors might have no discernible process, the vast majority of excellent actors have developed and continue to develop a set of tools that reliably works for them, role after role.

It might help to think about the sound-by-sound work actors do on dialects and accents. That also feels technical and constricting at first. You keep at it because you know it will help you tell the story with specificity and clarity. Eventually, the dialect itself becomes a deeply integrated part of your performance and something you think very little about moment-to-moment. Physical actioning is similar. Over time you'll discover that these concepts, and the tools developed from them, improve your spontaneity because you better understand what is happening when things aren't working and can diagnose it. It also liberates you to find joy in the most important mystery of the work: shared moments of discovery with partners made possible by your fully available self. Good performances are liberated and released precisely because of the care and detail during the process. Make the table first, then invite friends over for dinner.

While there are many mentions of research and neuroscience throughout the book, I want to end our time together focusing on a few items that I believe have the greatest potential to impact the way you think about acting and are most integrated into the physical actioning experiments. Rather than interrupt the arc of the experiments themselves, I've put the deeper dive into neuroscience here in the program notes. These ideas will be familiar from earlier. Knowing them should fortify you to keep exploring in embodied ways as you rehearse and tempt you to wonder more deeply about what might be next for us as theatre-makers.

These topics are:

1. Links between thought and action.
2. How verbs work in your brain.
3. The space around your body.
4. Physicalization and memory.
5. Prediction and repeating performances.

1. Thoughts Are Actions

It is easy to think of thinking and doing as independent processes. We all have thoughts or ideas that we don't "act" on. We all resist certain impulses because we fear the results, or change our minds, or were taught that they aren't acceptable. From our subjective experience, this might generate a belief that there is a wall between the thinking and the doing – some gate we choose to open and close at will. The truth is much more complicated, and science is still working to understand many of the ways our thoughts become actions. What we already know is that they are interrelated in important ways. This means that, as actors, your thoughts about what you're doing in the work can impact what your partner and the audience sees.

Mirror Neurons and Processes

If you've heard little else about neuroscience you may know about the idea of "mirror neurons." While many of the early experiments about this were done on primates, scientists are doing more and more work to understand how these processes work in the human brain. Some of this may help us understand the process of acting itself and some may be better suited to illuminating the experience of your audience. Essentially, the same regions of your brain that activate when you do an activity, also activate to varying degrees when you imagine doing that thing or see someone else doing it. For instance, I played the trombone as a kid. When I think about playing trombone right now, these neurons and pathways in my brain "light up." It is even possible my lips move a tiny bit toward the *embouchure* of playing, or my right-hand muscles tense a bit toward the shape they need to grasp the trombone's slide. In many cases this movement is invisible to the eye. The important point is that, in some ways, just imagining it is a degree of doing it.

Similarly, when I go to a concert and see a trombone player, some of the same areas are activated in my brain just by watching them play. Even more exciting is that, even though I've never played the violin, when my attention goes to that section of the orchestra, this same process happens. It seems clear to scientists that my response isn't as strong in me as it is when I watch the trombone section, but some part of me is still imagining and modeling how my body would do that activity. While there may be times that we are conscious of this process happening because of the size of the visible response in our body, the process itself happens below our level of conscious awareness.[1]

Various studies explore exactly why we do this. Some suspect it is part of the process by which we experience empathy – helping us put ourselves in the shoes of others. Others investigate the relationship to learning. Still others look at the predictive power of our minds, exploring how we create predictive models of what comes next. An important distinction can be made between the way this works in humans and in primates. Studies indicate that human mirror processes activate when seeing both transitive and non-transitive actions (like grabbing a piece of food versus shrugging your shoulders). This suggests an ability in humans to imagine the purpose and intention of behavior that doesn't have a clear object-oriented goal. It is, perhaps, that humans know that other humans behave in ways to accomplish something even if there isn't an obvious goal visible.[2] You know that your date's foot tapping might be a sign of restlessness or boredom, not just a tick.

This knowledge about mirror neurons has a few potential important lessons for you as an actor. If we know that the mind of your partner on stage is always seeing and working to understand a "why" behind your actions, and we know that even your thoughts about doing something begin an embodied process, then *everything you do and think has the potential to communicate to your partners*. Selecting clear actions moment-to-moment provides them with an exciting flood of specific information. Perhaps that seems obvious already, but we have all seen performances where the original impulse for the behavior or words is a distant memory and all that is left is the empty shell of the moment.

It's easy to assume that actions, speech, or behaviors devoid of a real impulse to accomplish a goal might communicate the same thing. They don't. As a 2006 *New York Times* article notes:

> In a study published in March 2005 in Public Library of Science, Dr. [Marco] Iacoboni [at the University of California, Los Angeles] and

> his colleagues reported that mirror neurons could discern if another person who was picking up a cup of tea planned to drink from it or clear it from the table.[3]

Your intention as you do the action communicates. Simply doing the behavior is not enough.

If this is true of your partners, it is also true of the audience. The more embodied and physically engaged your performance, the more your audience experiences it along with you – simulating your place in the events. Their own brains fire in an attempt to mirror the physical life of your character. The early work of Italian scientist Giacomo Rizzolatti demonstrated that "watching something is the same as doing something – the same neurons fire."[4] We now know that it is even more complex than that. Your mind (and that of your audience) doesn't just mirror what it sees, it also creates a predictive model of what it expects to see and what it expects to come next.

Imagine two characters sitting on a bench. A wants to hold B's hand. As A sits there and imagines over and over all the ways they might reach over for B's hand, those neurons are firing. Sitting there isn't enough. Playing the state of "nervous" isn't necessary or interesting. Could you, as an actor, become more attuned to the small muscular impulses that arise from just that thought? What action verb helps you think and activate those impulses? Could you become aware and specific enough that an audience's mirror neurons fire as they see or sense that impulse? You've probably done this many times before without realizing it, or have experienced it as an audience member. Even better to cultivate and harness it as a tool for your work. Using this knowledge requires action-oriented acting. It requires that you keep your work focused on the world outside you rather than your internal state.

Other Non-Verbal Cues

Part of the reason physical actioning focuses so much on developing non-verbal communication and linking it with the text is because so much of what we communicate to partners and the audience is non-verbal. If you've had a meal with someone you were in an argument with, then you know that not all communication is verbal. There is now clearer evidence that a significant amount of the information we receive while interacting with others is non-verbal. This goes far beyond the obvious examples above. Every conversation is full of non-verbal cues and information that we experience as listeners both consciously and unconsciously.

We get non-verbal information from small facial movements and breath patterns and eye movement and many other cues that we might not be consciously aware of. One example is from the work of scientist Paul Ekman. He and his colleagues developed a well-known system for analyzing facial movement called FACS (Facial Action Coding System). They describe it as a "comprehensive, anatomically based system for measuring all visually discernible facial movement."[5] Some of their experiments investigated "micro expressions," which happen so quickly they last just a fraction of a second.[6] Despite their speed, they are still visible to the naked eye.[7] Most of us don't know what to look for, but we still witness these brief glimpses of repressed emotion and register them as a part of non-verbal communication. These micro expressions are like a strong impulse for a physical action you want to play but the circumstance and rules of the world squeeze, reshape, and suppress. Some bit of that impulse cannot help but manifest somewhere in your body.

Embodied Speech

Importantly, we also understand that verbal communication has a physical component that is not just some extra add-on, but a fundamental part of using spoken language to communicate. Thought, gesture, and speech are linked. The path from inspiration to verbal communication goes through a physicalization process that includes far more of your body than just the parts directly responsible for speech sounds. In fact, gesture can be an important part of how speech and ideas come to be formed and how listeners understand the complexity of the ideas behind speech.[8] Areas of the brain related to language processing are also activated in sequencing movement.[9] Scientists continue to discover more and more about just how connected thought, speech, and movement are.

Knowing this deep integration, any way of working on a role must include not just a process of intellectually understanding the thoughts of a character that turn into the spoken text, but how those thoughts are embodied as a part of and beyond the verbal impulses. Resist separating the elements of verbal and non-verbal communication in the rehearsal process since they are inherently linked in your body. The wall we've built between them is artificial. Thus, any cohesive set of tools must find a way to discuss thought itself as an embodied idea. As a result, our tendency to start a rehearsal process with "table work" and then move to an embodied exploration of the spoken text reinforces ideas about a divide that we now know doesn't exist. In fact, moments of table work might benefit significantly and provide more insight into the text if we encouraged actors to engage their bodies beyond their voices from the very beginning.

"Movement Work" as Embodied Thought

Much of the movement training you receive as an actor focuses on one of a few aspects of the use of the body in performance. First, you become aware of and able to exclude parasitic physical habits from your life, so they are not present in all of your performances. This could be anything from a small habitual gesture to postural issues that might undermine your believability in certain roles. Perhaps there is a certain way you hold excess tension in your back or protect some part of yourself because of an old injury. Everyone has these habits and they're unique from person to person. From there, you develop good habits of self-use, so you are able to work in rehearsal and performance as needed while ensuring that you aren't harming yourself meeting the frequently substantial demands on your bodies. Finally, you learn techniques for building physicality that are specific to a given character and differentiates the perceived body of the character from you in large or small ways depending on the role. This is all vital work and contributes substantially to the expressive bodies of actors we experience in performance. These techniques also go a long way toward both liberating your ability to engage in non-verbal communication and toward crafting some of that communication in character-specific ways.

As you learned above, though, the audience experiences far more non-verbal communication than we teach you to express. You already work with rigor on the spoken version of the text – ensuring not a single word is dropped and doing sound-by-sound work to ensure that an accent or dialect is believable. We use a far less specific set of tools to ensure that non-verbal communication is just as clear and supports (or undermines) the text as needed for the narrative. This doesn't require planning every gesture in advance any more than voice work means planning every vocal inflection.

The next step in your physical work is taking your flexible body and training it to communicate with specificity and clarity moment-to-moment in performance. This means moving beyond character physicality, blocking, and period- or culturally specific social behavior to explore and harness all the information you share non-verbally with as much detail and specificity as possible. As you have discovered throughout this book, this includes linking your physical impulses directly to the text. It requires physical work that primes you to use non-verbal communication as richly and rigorously as verbal communication since both are connected at the impulse in your brain. Importantly, it means describing your impulses with a unified vocabulary, so every element of the expression is part of the whole.

2. Thinking Verbs

Since you know that thoughts can begin an engagement in the rest of your body and that partners and audiences receive a massive amount of information non-verbally based on your thoughts and impulses, *what* you think as an actor is an important piece of your communication. Throughout the book we used transitive verbs as the core way to speak about what you're doing moment-to-moment. This isn't a new idea, but often it's relegated to table work or script analysis, and then mostly left behind. Sometimes the way we phrase actions or objectives in acting techniques make them passive, self-focused, or too complex to accomplish in a moment, or gesture, or single sentence. In contrast, physical work around character gets decoupled from the exploration of text or moment-to-moment behavior, uses an entirely distinct vocabulary, and is interested in broader questions about self-use or "types." My goal is to unify the language for the actions you play through speech, behavior, activities and other movements with a vocabulary that we know uniquely impacts your brain and strengthens your connection to embodiment.

We have a growing understanding of what action verbs specifically do in our brains, how that relates to embodiment, and how deeply the associations with those words run. Evidence suggests that merely *thinking* these words begins a process of embodiment. One study showed that thinking about verbs, both very active or less active ("stabs" and "observes" might be examples), activated areas of the brain once attributed with the activation of visual-motion features. The research indicates that these verbs actually produce far more than merely a sensorimotor response. Rather, they generate a series of "event concepts" related to doing, remembering doing, and memories of doing the verb.[10]

Fascinatingly, our sense of these action verbs is not altered based on the way we learn them. In one study researchers looked at individuals who are sighted or congenitally blind to compare their processing of action verbs. They found that the same region activated whether the subject was or wasn't blind.[11] By starting your journey with your work on a role by thinking in physical terms, your mind is already generating a sense of the space and action related to the moment-to-moment events of the performance.

While it is easy to understand the way thinking about a primary verb like "hugs" engages the brain, there is also evidence that abstract action words like "tempts," for instance, engage some of these regions.[12] Earlier you explored the use of various kinds of verbs (primary, conceptual, and metaphorical) described by researchers Lakoff and Johnson. Their work reveals how much of our language and ideas have a spatial and movement component. Their linguistic work with metaphors uncovered deep connections between our thoughts about verbs and how we apply them to humans.

This is a valuable tool when your work as an actor does not allow for much visible physicality. Using the language of transitive verbs as the core vocabulary of your work applies a sense that motion and action are aimed at your partner even when you cannot take visible physical action. Your work in expanding and contracting the embodied form gives you the opportunity to experience a released version of actions, even if the Rules of the World and the production require a tremendously restrained performance. There is evidence that doing actions excites the same parts of the brain as imagining and observing those actions.[13] After exploring with partners through physical actioning, you enter performance with a deep experience of all of those versions to recall and use in the moment. The ultimate goal of the actor is *doing* on stage. The way you think about what you're doing to your partner benefits from putting embodied language at the center.

3. Peripersonal Space

So, if your thoughts have embodied impulses and we are now using a language that supports embodiment from thought through behavior and language, then the next step is to make sure these impulses become visible in a way that you and your partner experience clearly. You want to see your actions have a result on your partner. Our earlier discussion on mirror neurons describes some of what happens in your brain as you watch your partner and how that might spark motor systems, but when you take visible physical action other factors of perception also come into play.

Many of the exercises in this book ask you to explore work with partners in space, keep your focus on them, and to expansively physicalize actions and responses. Some of what is happening in these exercises relates to your peripersonal space. Peripersonal space is the area immediately around your body. In the exercises where you use "thick space" or expanded physical actions to alter your partner, you create a full body experience of how actions impact your peripersonal space. You and your partner engage in this imaginative environment where all physical movement has a direct physical impact on people, even when you aren't touching. You move your body in an attempt to play an action. Your partner allows that action to change the shape of their body in some way. That turns into a new action from them that changes you.

Recent evidence suggests that your body schema (sense of your own body) and peripersonal space change, based on your behavior and your expectations of how that behavior will alter your immediate environment. They also appear to be flexible in size, allowing for an expansion or contraction through the use of real or intangible tools – like stretching out your arms or making a sound. [14] Additional evidence reveals that your peripersonal space isn't a single unified zone, but rather it is specific to circumstances in your surroundings and how you are using your body.[15] It isn't the same in an open field and on a crowded subway car.

The exercise of playing actions through space and seeing the physical response from your partner is partly a process of testing your expectations about how effective your body was and how the result in their body met your expectations. You then change your action (or the way you play the current action) to refine and better meet your goal. It is easier to trick yourself into thinking your action succeeded if it's a planned line of text followed by your partner's planned line of text. When these actions take a more visible form (with or without the spoken text) you have undeniable visual feedback about the success or failure of that action. As you then contract the action back down to something more "realistic," you and your partner stay aware of the visible physical remnants of the expanded action and how it impacts your partner. As a result, you take this kinesthetic experience of a scene or moment into rehearsal, something you would not have without trying to engage a partner through your peripersonal space and physically alter them. Importantly, because you use a simple transitive verb to describe the thought, intention behind the spoken language, and any physical behavior/activity/movement, you have a singular embodied idea behind everything that is happening. They all unify to change your partner.

4. Physicalization and Memory

You know that much of what makes a character who they are will have happened before the text begins or out of the sight of the audience. Some of those events are joyful while others are terrible and traumatic. Some actors have a regrettably wide array of difficult experiences in their own past. Others do not. The demand to use your own life's difficulties in your work (or a feeling of inadequacy for not having more personal tragedy to draw on) has been a frustrating feature of actor training for many decades. Various acting techniques operate from strong views about the use of imagination versus/and personal memory.

It turns out that this major point of tension between different acting techniques might be far less important than many people make it. In fact, the act of remembering is, in large part, an act of imagination and is impacted significantly by the situation in which we are attempting to remember. You do not reach onto a metaphorical shelf and pull down that memory. You create it from bits of memory influenced by the larger circumstances of the original event, the other times you've remembered it, and the current situation.[16] You likely have your own experiences that reveal how faulty memory can be. Perhaps you vividly remember a special event and the dress you wore, only to see a picture and realize that that dress was worn at an entirely different event. Certainly, you've heard a friend or family member describe an event you witnessed only to think that their description is entirely wrong. I have a distinct memory of being just three months old and sitting in a high chair in a house where my family lived. There's just one problem – that's not where we lived when I was a baby.

Remembering, imagining, and predicting how future events will go are related processes in the brain.[17] It is easy to understand why, intuitively, acting teachers so often ask students to remember related personal events that might help them produce a performance aligned with the imaginary circumstances of the text. Having particularly traumatic or varied life events to draw from as you confront circumstances for a character might be useful, but it isn't necessary. This is good news for any actor who fears their lack of similarity to a character's life or relative lack of life experiences makes them incapable of playing a role.

It also may help to explain the resistance many actors have to imagining terrible events in their character's past or doing exercises where they imagine horrible things happening to people they know. One MIT study demonstrated that imagining a face and imagining a location activated specific regions in the brain related to perceiving those items when they're present.[18] Your imagined engagement and actual perception are deeply related. Imagining that you witness someone being murdered isn't the same thing as actually witnessing it, but with thorough and informed imaginative exploration your brain will have many similar responses. In practice, it means that you can use exercises that allow both memory and imagination to play a part and liberate you from believing the "realness" of your memories somehow stands above the "falseness" of your imagination.

In Part 2 you explored this link between memory and imagination by embodying events from your life noted in your Physical Action Journal and reproducing them in various ways. We know that physicalizing aspects of the events that you are trying to remember can help you recall the events.[19] We all have the experience of walking around retracing our steps in an attempt to recall where we left something. Perhaps you also recognize how you shift your body position or face or re-enact gestures when recalling a conversation or event. These behaviors reflect an intuitive understanding that your body can help you strengthen your ability to recall moments from the past.

In Part 3 you explored some parallel exercises with events from the text or imagined events from the life of the role you were playing – walking a number of events to build embodied experiences you can draw on as memories in later rehearsal and during performances. Physicalizing these imagined events as you rehearse builds synaptic connections that fire when you later go to perform those moments.[20] It is great to write or even discuss in rehearsal an event that your character experienced as a child, for instance. Even better to act it out. When you later speak about it in the play, neurons fire and your body re-experiences some part of that imagined event.

This relationship between physical experience and both recalling and generating memories is part of why you benefit from the exercises earlier in the book. Taking the physical shape and moving around in the way you recall from an actual event helps you remember. The embodied playing of the imaginary events helps build more connections in your brain related to those events. Then, when you come to perform, you can briefly physicalize a moment from that real or imagined event to recall that memory in a more deeply connected way in front of the audience.

5. Predictive Processing

Acting and building a performance include a great deal of spontaneity and discovery. The act of performing itself, however, requires tremendous repetition of the same words and specifically timed physical events. Your ability to generate seeming spontaneity despite these requirements is challenged by what we now understand about how our brains work.

Our brains are interested in assessing danger and focusing on what is important while tuning out what we think is noise or unimportant detail. We are on the lookout for what might cause us harm and, as a result, what can be ignored. One of the results of this is a phenomenon known as "change blindness," where we miss major alterations in a scene or event because they are not the element we think needs our attention.[21] Some amusing examples include switching out who someone is speaking to or having enormous events happen in plain view while the subject remains totally unaware of the change. Evidence suggests that part of the way we keep focus on what seems important is through prediction – generating models of spaces, and people, and expectations about what exists, which feed our perception. It is only when we notice that the stimuli from the environment does not match that predictive model that we start to pay attention in a conscious way to reassess. Instead of being a fixed system where perception and prediction are always working in the same way, it seems fluid and altered by attention.[22]

As you can imagine, this natural and valuable way our brains work in life creates a challenging scenario for actors. The rehearsal process generates tremendous consistency over time in the performances of you and your partners. It must, in order to be a repeatable event, meet the technical demands of the larger production, and allow your collaborators to do their work. Knowing this, you should work to both generate and notice small differences as part of your performance technique. I don't mean simply making arbitrary changes each performance. You must repeat the performance with the detail you and your director worked to achieve while also finding new things that do not fundamentally change the performance.

This requires that you have clarity about what you are doing moment-to-moment and how it serves the project and story without a rigid consistency of embodied expression. It may also mean adherence to incredibly specific physical demands (e.g. take a drag on the cigarette after a particular word) without those activities becoming hollow or robotic over time. How do you manage to balance these demands? One key component is focusing on the source of the behavior you must perform through the underlying action rather than the visible behavior itself. If the result that your partner and audience see is a score of physical actions shaped by the Rules of the World, then you can continue to do the same action performance after performance, often even using the same behavior or activities, and discover small spontaneous changes and adjustments that keep the integrity of your performance. Focus on the impact that smoking the cigarette should have on your partner and watch for that result rather than focusing on the cigarette or your experience of smoking it.

The same is true when you watch your partners in performance. Much of what happens after several rehearsals is predictable. You must practice seeking the difference, not to overvalue it, but to ensure that you stay present and avoid slipping into an unconscious ride of predictive experience. By shifting your curiosity and attention to what is different *this time* you help rebalance your natural predictive impulses and reduce how much becomes background noise. These observations in the moment inspire your small shifts, providing something new for your partner to observe. The more you and your partner know the predictive nature of your brains, the more you can help one another see and generate new information in each rehearsal and performance.

Scaring people on stage is a great example of this challenge. It's very hard to scare a fellow actor or be scared repeatedly. A real reaction of fear is so recognizable and a false one so much less satisfying. Many actors overcome this by trying to scare their partner slightly differently every time or by altering the timing just a bit. Sometimes the actor receiving the scare distracts themselves by putting attention elsewhere in the moments before the scare. All of these methods intuitively understand that anticipation of the surprise makes it fail. As soon as you can predict it is coming, the real impulse is gone. Those adjustments are some of the same ones required for any moment-to-moment work. How are you setting your partner up to help them succeed in telling the story? How can you make your action consistent and reliable for the storytelling but new and discovered and shaped to help your partner avoid prediction?

Your Well-Made Shoes

Most actors intuitively understand that acting is a physical craft. It wasn't until recently, though, that we understood just *how* physical a job it is. Even the imaginative process is embodied. This new knowledge demands that we develop new tools that eliminate the language and assumptions about this invented mind/body divide and leave it behind for good.

The experiments throughout this book incorporate and theorize using emerging knowledge about neuroscience with an understanding of the actor's process and how an actor must work in a collaborative environment. The experiments seek to deeply integrate embodiment into your thoughts about character and moment-to-moment interaction, generate physical experiences of events to help you build memory and connections, and provide a kinesthetic experience of the interaction between partners to illuminate the sometimes hidden impulses, and they demand that you keep looking for what is new and changing to keep things specific and reliable, but always alive.

My hope is that these experiments provide you with a path to rethink your physical work as an actor. Let go of the idea that embodiment in your process is the abandonment of thought or "getting out of your head." Let go of the idea that thinking about and discussing a script is a step before your body becomes involved – it is already involved while you are thinking. If you abandon those deeply reinforced beliefs, then you must reconsider the shape of your process and all the ways those assumptions guide it. Use embodiment from the start. Enact events so that thinking of them later sparks that embodied memory in your brain. Explore how your behavior and physical life relates to these impulses. Think of speaking as a form of embodiment. Develop tools that reveal and strengthen your physical impulses, while acknowledging how the Rules of the World shape them. Finally, embrace our new understanding about how your body (including your brain) works, just as actors embraced psychology and used it to better represent humanness in the last century. Will your brain get in your way? Your brain already *is* your way, so use that to your advantage as you work.

You can now begin your journey with each role using the language of action, tools that harness those impulses, and confidence in your capacity to handle the restrictions and liberations of any project or venue. You know more than you did when you began. You can tell by all the pages in your left hand versus your right that we're almost done. Still, as you close this book, I hope you'll return immediately to the first task again. Lift your eyes with curiosity.

Notes

1 Garner, Stanton B. Jr. *Kinesthetic Spectatorship in Theatre: Phenomenology, Cognition, Movement*, Cognitive Studies in Literature and Performance. Cham, Switzerland: Palgrave Macmillan, 2018, pp. 156–157. DOI: 10.1007/978-3-319-91794-8.
2 DiPaolo, Ezequiel and Evan Thompson. "The Enactive Approach." In *The Routledge Handbook of Embodied Cognition*, edited by Lawrence Shapiro. New York: Routledge, 2014, p. 55.
3 Blakeslee, Sandra. "Cells that Read Minds." *The New York Times*, January 10, 2006; June 15, 2019. https://www.nytimes.com/2006/01/10/science/cells-that-read-minds.html.
4 Blair, Rhonda. *The Actor Image and Action: Acting and Cognitive Neuroscience*. New York: Routledge, 2008, p. 13.
5 Rosenberg, Erika. "Introduction: The Study of Spontaneous Facial Expressions in Psychology." In *What the Face Reveals: Basic and Applied Studies of Spontaneous Expression Using the Facial Action Encoding System (FACS)*, edited by Erika L. Rosenberg and Paul Ekman. New York: Oxford University Press, 1997, p. 13.
6 Ekman, Paul. *Emotions Revealed*. New York: Holt and Company, 2007, p. 214.
7 Ibid., p. 215.
8 Goldin-Meadow, Susan. "The Role of Gesture in Communication and Thinking." *Trends in Cognitive Science* 3, no. 11 (1999): 419–429. http://citeseerx.ist.psu.edu/viewdoc/download?doi=10.1.1.114.4949&rep=rep1&type=pdf.
9 Lutterbie, John. *Towards a General Theory of Acting*, Cognitive Studies in Literature and Performance. New York: Palgrave Macmillan, 2011, p. 124. DOI 10.1057/9780230119468.
10 Bedny, Marina, Alfonso Caramazza, Emily Grossman, Alvaro Pascual-Leone, and Rebecca Saxe. "Concepts Are More than Percepts: The Case of Action Verbs." *The Journal of Neuroscience* 28, no. 44 (2008): 11347–11353. DOI: 10.1523/JNEUROSCI.3039-08.2008.
11 Bedny, Marina, Alfonso Caramazza, Alvaro Pascual-Leone, and Rebecca Saxe. "Typical Neural Representations of Action Verbs Develop without Vision." *Cerebral Cortex* 22 (2012): 286–293. DOI:10.1093/cercor/bhr081.
12 Sakreida, Katrin, Claudia Scorolli, Mareike M. Menz, Stefan Heim, Anna M. Borghi, and Ferdinand Binkofski. "Are Abstract Action Words Embodied? An fMRI Investigation at the Interface between Language and Motor Cognition." *Frontiers in Human Neuroscience* 7 (2013): 1–13. DOI: 10.3389/fnhum.2013.00125.
13 Savaki, Helen E. and Vassilis Raos. "Action Perception and Motor Imagery: Mental Practice of Action." *Progress in Neurobiology* 175 (2019): 107–125. DOI: 10.1016/j.pneurobio.2019.01.007.

14 D'Angelo, Mariano, Giuseppe di Pellegrino, et al. "The Sense of Agency Shapes Body Schema and Peripersonal Space." *Scientific Reports* 8 (2018): 13847. DOI: 10.1038/s41598-018-32238-z.
15 Bufacchi, Rory J. and Gian Domenico Iannetti. "An Action Field Theory of Peripersonal Space." *Trends in Cognitive Sciences* 22, no. 12 (2018): 1076–1090. DOI: 10.1016/j.tics.2018.09.004.
16 Kemp, Rick. *Embodied Acting: What Neuroscience Tells Us about Performance.* New York: Routledge, 2012, p. 161.
17 Mullally, Sinéad L. and Eleanor A. Maguire. "Memory, Imagination, and Predicting the Future: A Common Brain Mechanism?" *The Neuroscientist* 20, no. 3 (2014): 220–234. DOI: 10.1177/1073858413495091.
18 O'Craven, K.M. and N. Kanwisher. "Mental Imagery of Faces and Places Activates Corresponding Stimulus-Specific Brain Regions." *Journal of Cognitive Neuroscience* 12, no. 6 (2000): 1013–1023.
19 Dijkstra, Katina and Rolf A. Swan. "Memory and Action." In *The Routledge Handbook of Embodied Cognition*, edited by Lawrence Shapiro. New York: Routledge, 2014, pp. 296–305.
20 Sutton, John and Kellie Williamson. "Embodied Remembering." In *The Routledge Handbook of Embodied Cognition*, edited by Lawrence Shapiro. New York: Routledge, 2014, pp. 315–325.
21 Tversky, Barbara. *Mind in Motion: How Action Shapes Thought.* New York: Basic Books, 2019. 50–51.
22 Clark, Andy. *Surfing Uncertainty: Prediction, Action, and the Embodied Mind.* Oxford: Oxford University Press, 2015, p. 57. DOI: 10.1093/acprof:oso/9780190217013.001.0001.

Appendix

Action Verb List: Alphabetical Order

A

Abandons
Absolves
Accepts
Accuses
Admonishes
Adores
Agitates
Aids
Alarms
Amazes
Anchors
Annihilates
Avoids

B

Baits
Banishes
Bashes
Beckons
Belittles
Bends
Betrays
Bites
Blasts
Blesses
Bludgeons
Boosts
Bribes

C

Caresses
Charms
Chills
Coaxes
Coddles
Condemns
Conquers
Corners
Courts
Cuddles
Cultivates
Cures
Curses

D

Dazzles
Defends
Demolishes
Deserts
Devours
Disarms
Disciplines
Disgraces
Dismisses
Disorients
Dissects
Drives

E

Eases
Educates
Elbows
Electrifies
Elevates
Embarrasses
Emboldens
Embraces
Enchants
Endows
Enlightens
Enlists
Entertains
Entices
Escorts
Extinguishes

F

Fans
Feeds
Flicks
Forgives
Freezes
Fuels

G

Grasps
Guards
Guides

H

Halts
Heals
Hoists
Honors
Hooks
Hugs
Humiliates
Hypnotizes

I

Ignites
Impales
Incites
Infects
Inflates
Inspects
Interrogates
Invites

J

Jabs
Joins

Jolts
Jostles

K

Kicks
Kindles
Kisses

L

Liberates
Licks
Lures

M

Maims
Massages
Melts

Mirrors
Mocks
Mothers

N

Nabs
Nourishes
Nudges
Nurses
Nuzzles

O

Obeys
Offends
Overpowers

P

Pardons
Parents
Patronizes
Permits
Pets
Pins
Pleasures
Poisons
Pokes
Presses
Prompts
Propels
Provokes
Pulls
Purifies
Pushes

Q

Quizzes

R

Rallies
Rebuilds
Redirects
Rejects
Rewards
Roots
Rushes

S

Sabotages
Saddles
Salutes
Saps
Satisfies
Scans
Scrubs
Seduces
Shakes
Sharpens
Shelters
Shepherds
Shocks
Shuns
Silences
Skewers
Slams
Slaps
Slashes
Smoothes
Smothers
Snows
Soothes
Squeezes
Stabilizes
Stabs
Stalls
Steadies
Steers
Stings
Stokes
Stops
Studies
Subdues
Surprises
Swats

T

Tackles
Taints
Taunts
Teaches
Teases
Tempts
Thaws
Thrills
Throttles
Thwarts
Tickles
Tolerates
Treasures
Tugs

U

Unleashes
Uplifts

V

Vilifies

W

Warms
Warns
Welcomes
Worries
Worships
Wounds

Y

Yanks

Action Verb List: By Type

PRIMARY ACTIONS
Bashes
Bends
Bites
Blasts
Bludgeons
Caresses
Corners
Cuddles
Elbows
Embraces
Fans
Feeds
Flicks
Grasps
Halts
Hoists
Hugs
Impales
Jabs
Jostles
Kicks
Kisses
Licks
Maims
Massages
Mirrors
Nudges
Nuzzles
Pets
Pins
Pokes
Pulls
Pushes
Rushes
Salutes
Scans
Scrubs
Shakes
Slams
Slaps
Slashes
Smothers
Squeezes
Stabs
Swats
Tackles
Tickles
Tugs
Wounds
Yanks

CONCEPTUAL ACTIONS
Abandons
Absolves
Accepts
Accuses
Admonishes
Adores
Agitates
Aids
Alarms
Amazes
Annihilates
Avoids
Banishes
Beckons
Belittles
Blesses
Betrays
Bribes
Charms
Coaxes
Coddles
Condemns
Conquers
Courts
Cures
Curses
Dazzles
Defends
Deserts
Disciplines
Disgraces
Dismisses
Disorients
Eases
Educates
Embarrasses
Emboldens
Enchants
Endows
Escorts
Enlists
Entertains
Entices
Forgives
Guards
Guides
Heals
Honors
Humiliates
Hypnotizes
Incites
Inspects
Interrogates
Invites
Joins
Jolts
Liberates
Mocks
Mothers
Nabs
Nurses
Obeys
Offends
Overpowers
Pardons
Parents
Patronizes
Permits
Pleasures
Prompts
Provokes
Quizzes
Rallies

Redirects
Rejects
Rewards
Sabotages
Satisfies
Seduces
Shuns
Silences
Soothes
Steadies
Stops
Studies
Subdues
Surprises
Taunts
Teaches
Teases
Tempts
Thrills
Thwarts
Tolerates
Treasures
Vilifies
Warns
Welcomes
Worries
Worships

METAPHORICAL ACTIONS

Anchors
Baits
Boosts
Chills
Cultivates
Demolishes
Devours
Disarms
Dissects
Drives
Electrifies
Elevates
Enlightens
Extinguishes
Freezes
Fuels
Hooks
Ignites
Infects
Inflates
Kindles
Lures
Melts
Nourishes
Poisons
Presses
Propels
Purifies
Rebuilds
Roots
Saddles
Saps
Sharpens
Shelters
Shepherds
Shocks
Skewers
Smoothes
Smothers
Snows
Stabilizes
Stalls
Steers
Stings
Stokes
Taints
Thaws
Throttles
Unleashes
Uplifts

Exercises

Index

Note: Page numbers in **bold** refer to the individual exercises